MOTHER'S AROMA
JIN YUANSHAN PATCHWORK ARTS

BY JIN YUANSHAN

金媛善 著

母亲的香气
——金媛善拼布艺术

国家一级出版社
中国纺织出版社
全国百佳图书出版单位

内 容 提 要

本书详细生动地介绍了拼布艺术的起源以及朝鲜族和中国少数民族的拼布艺术,重点向读者详细讲解了拼布钱形纹和太阳花的制作,并展示了金媛善大师拼布艺术作品家居类(39幅)、服饰类(44幅)、壁画类(14幅)、包袱皮(102幅),精妙的配色以及设计给读者展现了拼布的独特魅力。它们的共同主题是"女性""家人""祝福""关怀""勇气""坚持"和"责任"。作者认为,这是能让他人幸福,也能给自己带来幸运的最重要的东西。

本书适合服装设计专业师生、拼布艺术爱好者以及从事艺术创作的专业人士使用。

图书在版编目(CIP)数据

母亲的香气:金媛善拼布艺术/金媛善著. —北京:中国纺织出版社,2019.12
ISBN 978-7-5180-5706-1

Ⅰ.①母… Ⅱ.①金… Ⅲ.①布料-手工艺品-制作 Ⅳ.① TS973.5

中国版本图书馆 CIP 数据核字(2018)第 280694 号

责任编辑:谢冰雁　　责任校对:寇晨晨
责任设计:何　建　　责任印制:王艳丽

中国纺织出版社出版发行
地址:北京市朝阳区百子湾东里 A407 号楼　邮政编码:100124
销售电话:010 — 67004422　传真:010 — 87155801
http://www.c-textilep.com
中国纺织出版社天猫旗舰店
官方微博 http://weibo.com/2119887771
北京华联印刷有限公司印刷　各地新华书店经销
2019 年 12 月第 1 版第 1 次印刷
开本:787×1092　1/8　印张:35.5
字数:213 千字　定价:258.00 元

凡购本书,如有缺页、倒页、脱页,由本社图书营销中心调换

金媛善

朝鲜族,卓越的拼布艺术家。1947年5月出生,哈尔滨人,自幼受祖母、母亲的熏陶,擅长女红。20世纪90年代开始遍访各少数民族地区,考察、学习各民族手工传统技艺。数十年如一日执着于拼布艺术的研究和创作。现任中国流行色协会拼布色彩与艺术研究专业委员会副主任、首席专家,中国流行色协会荣誉理事,北京服装学院民族服饰博物馆专家顾问,北京服装学院天工传习馆荣誉专家。北京服装学院天工传习馆金媛善拼布工作室创始人。

多次参加中国、日本、韩国、美国和加拿大的拼布艺术展览,2007年获国际拼布二等奖,2008年获国内银奖。

2003年、2006年、2009年应邀在韩国举办个人作品展。作品先后被韩国、日本、澳大利亚、美国、加拿大等国家的艺术机构和个人收藏。

2009年3月应邀在清华大学美术学院举办个人作品展并发表论文,同时举行拼布艺术教学活动。

2010年10月应邀在北京服装学院举办个展并举行拼布艺术教学活动。

2011年3月在清华大学美术学院举办的"2011年第十一届全国纺织品设计大赛暨国际理论研讨会"上演讲。

2012年8月受邀以拼布作家的身份参加中国台湾国际艺术拼布节并做演讲;同时举行教学活动。

2012年9月被联合国教科文组织授予"杰出手工品认证徽章",并作品收藏。

2012年当选世界艺术拼布作家协会(Studio Art Quilt Associates)中国区代表。

2012年12月获得"中国色彩事业推动奖"。

2014年6月金媛善受到文化部邀请,随团参加在美国华盛顿举办的第48届史密森尼民俗节,这是我国首次以官方名义参加该活动。

2014年获得"2013/2014年度中国色彩文化奖",同时获得"流行色协会第八届理事会先进工作者"和"中国流行色协会荣誉理事"两项殊荣。

2015年在美国哥伦比亚大学、耶鲁大学等五所高校及联合国总部展览。

2017年荣获色彩中国2017"中国传统工艺色彩奖(拼布类)"。

2017年被评为"非凡时尚人物"。

2018年受聘北京理工大学珠海学院设计与艺术学院客座教授。

在当今国内拼布复制欧美纹样盛行的潮流中,金媛善女士以其巧夺天工的作品独树一帜,让世人看到了中国拼布艺术的独特风貌和沉静典雅、变幻无穷的魅力。

Jin Yuanshan

Korean nationality, an excellent patchwork artist. She was born in May 1947 in Harbin and educated by her grandmother and mother since childhood, good at needlework. In the 1990s, she began to visit various ethnic minority areas to inspect and study the traditional craftsmanship of various ethnic groups. For decades, she has been obsessed with the research and creation of patchwork art. She is currently the deputy director and chief expert of Patchwork Color and Art Research Committee of China Fashion & Color Association, the honorary director of the China Fashion & Color Association, the expert consultant of the National Costume Museum of the Beijing Institute of Fashion Technology, and the honorary expert of the Tiangong Inheritance Culture Institute of the Beijing Institute of Fashion Technology. Founder of Jinyuanshan Patchwork Studio, Tiangong Inheritance Culture Institute of the Beijing Institute of Fashion Technology.

She has participated in many patchwork exhibitions in China, Japan, Korea and the United States. In 2007, she won the second prize of International Patchwork and won the domestic silver medal in 2008.

In 2003, 2006 and 2009, she was invited to hold a solo exhibition in Korea. Her works have been collected by art institutions and individuals from Korea, Japan, Australia, the United States, Canada and other countries.

In March 2009, she was invited to hold a solo exhibition and publish a thesis at the Academy of Fine Arts of Tsinghua University, and at the same time teach the art of patchwork.

In October 2010, she was invited to hold a solo exhibition at the Beijing Institute of Fashion Technology and to teach the art of patchwork.

In March 2011, she gave a speech at the 11th National Textile Design Competition and International Theory Seminar held at the Academy of Fine Arts of Tsinghua University.

In August 2012, she was invited to participate in the China Taiwan International Art Patchwork Festival as a patchwork writer and gave a speech.

In September 2012, she was awarded the "Outstanding Handicraft Certification Badge" by UNESCO.

In 2012, she was elected as the representative of Studio Art Quilt Associates in China.

In December 2012, she won the "China Color Career Promotion Award".

In June 2014, Jin Yuanshan was invited by the Ministry of Culture to participate in the 48th anniversary of the Smithsonian Folk Festival held in Washington, DC. This is the first time that China officially participated in the event.

In 2014, she won the "2013/2014 China Color Culture Award", and "Advanced Worker of the 8th Council of the China Fashion & Color Association" and "Honorary Director of the China Fashion & Color Association".

In 2015, she exhibited at five universities including Columbia University and Yale University and the United Nations Headquarters.

In 2017, she won the "Chinese Traditional Craft Color Award (Pencil)" of Color China 2017.

In 2017, she was named "Extraordinary Fashion Icon".

In 2018, she was invited as a visiting professor at the School of Design and Art, Zhuhai College, Beijing Institute of Technology.

Today, reproduction of European and American patterns are prevailing in domestic patchwork, Ms. Jin Yuanshan's work is unique and outstanding, allowing people see the unique style of Chinese patchwork art and the charm of quiet, elegant and ever-changing. Ms. Jin Yuanshan's work has been highly recognized by many experts and scholars, and she is hailed as "the only artist who can represent the art level of Chinese hand-made patchwork."

序 1

金媛善女士是当今中国乃至世界闻名的杰出拼布艺术家,作为北京服装学院民族服饰博物馆的专家顾问,金女士一直关注并从中国民族、民间传统手工艺术中汲取营养,最终逐渐形成了她独特的艺术风格。2014年,金媛善拼布工作室入驻北京服装学院时尚产业创新园天工传习馆,使得校内师生和广大拼布爱好者有更多机会了解和学习这门古老而又时尚的手工艺术,并使传统手工艺融入现代生活,成为现代服装、家居纺织品设计的重要手法,令拼布艺术焕发出新的时代风貌,这与北京服装学院历来倡导的民族服饰文化的保护传承创新宗旨是一脉相承的。

从金媛善女士的作品中,我们可以深深感受到中华优秀传统文化的积淀。金媛善女士从20世纪90年代便开始走访中国各少数民族聚居地区,考察和学习各民族传统的手工技艺,并在学习传统的基础上进行创新发展。例如,金女士从苗族、彝族、壮族、满族、藏族、土族和朝鲜族等各民族的传统手工艺中,学习拼布艺术中常用的拼缝、贴布、叠布等方法,并将少数民族拼布中的"倒三针"针法运用到作品之中,体现出中国拼布艺术所特有的本土化风格。因此,每当金媛善女士的拼布作品在美国、加拿大、韩国、日本等国际拼布艺术展览上亮相时,总会因为其鲜明的中国艺术语言而受人瞩目和喜爱。

金媛善女士对拼布艺术创作的热忱还植根于内心对生命和美好生活的憧憬与向往。书中收录的许多作品不仅具有审美性,更注重实用性,如做给丈夫的棉被、给儿子的窗帘和桌旗、给孙子孙女的"百天衣"和"周岁服"。通过这些拼布作品,我们可以细细体味到金女士在丈夫被病魔缠身时的祈愿;在儿女结婚时的祝福;在孙子孙女出生和周岁时的希冀……在这些作品中,每一件都投入其真情实感,无一不渗透着金媛善女士对生命和生活的热爱,以及要把这种浓浓的爱意传播给身边亲人的执着信念。她和她的作品集中体现出中国女性沉静平和、朴素坚韧、慧心巧思的优秀品质和传统美德。

相信这本书的出版一定能够引发更多的人来关注和喜爱拼布艺术,以及它背后的中国传统文化与艺术的真谛。

是为序。

北京服装学院 教授 刘元风
2019年夏

Prologue 1

Mrs. Jin Yuanshan is an outstanding patchwork artist in China and even in the world. As an expert consultant of the Ethnic Costume Museum of Beijing Institute of Fashion Technology, Mrs. Jin has been paying attention to and drawing nutrition from Chinese ethnic and folk traditional handicrafts, and gradually formed her unique art style. The Jin yuanshan Patchwork Studio was established in 2014 in Tiangong Inheritance Culture Institution, which located at the Fashion Industry Park of Beijing Institute of Fashion Technology, allowing the teachers, students and patchwork lovers have more opportunities to learn this ancient and fashionable art. Integrating traditional handicrafts into modern life and making it an important method in the design of modern clothing and home textiles, brings the art of patchwork to a new era. This is in line with the concept on the ethnic costume culture of protection, inheritance and innovation advocated by the Beijing Institute of Fashion Technology.

From the works of Mrs. Jin Yuanshan, we will appreciate the accumulation of Chinese excellent traditional culture. Since the 1990s, Mrs. Jin Yuanshan has visited various ethnic minority areas in China to investigate and study the traditional craftsmanship of various ethnic groups, so as to innovate and develop it on the basis of tradition. For example, Mrs. Jin learned the methods of quilting, patching, stacking, etc. commonly used in patchwork art from the traditional handicrafts of the Miao, Yi, Zhuang, Manchu, Tibetan, Tu, and Korean ethnic groups. The "reverse three stitches" method used in the patchwork in ethnic groups is applied, reflecting the unique local style of Chinese patchwork art. Therefore, whenever Mrs. Jin Yuanshan's patchwork are displayed in international art exhibitions in the United States, Canada, South Korea and Japan, they always attracts people's attention due to their distinctive Chinese art language.

Mrs. Jin Yuanshan's enthusiasm for patchwork art creation is rooted in her inner longing and yearning for a good and beautiful life. Many of her works included in the book are not only aesthetic, but also practical, such as quilts for her husband, curtains and table runners for her children, Baitianyi (100-Day Garment) and Zhousuifu (One Year Old Garment) for her grandchildren. Through these patchworks, we could understand the hope of Mrs. Jin when her husband was seriously illness; the blessing for her children when they got married: the concern for her grandchildren when they were born and one year old... Each of these works reflects Mrs. Jin Yuanshan's true feelings, deep love for life, and her persistent belief in spreading this deep love to her beloved ones. She and her works embody the excellent qualities and traditional virtues of Chinese women, they are calm and peaceful, simple and tough, smart and ingenious.

I believe that the publication of this book will definitely attract more people to pay attention to and love the art of patchwork, as well as the essence of Chinese traditional culture and art behind it.

That's all.

Professor Liu Yuanfeng
Summer, 2019

序 2

拼布是将一块块布料缝制起来的一种技艺，针线游走于不同形状面料之间，用色彩和图案做成作品，有着不同的用途和寓意，传递美的感受。拼布属于女红，我国56个民族中有33个民族都具备拼布技艺，包含28种以上的手工针法，可以说我国少数民族拼布图案涵盖了流行的拼布图案。传统拼布艺术起源于中国，在大英博物馆中收藏了一幅来自中国唐朝的拼布艺术品。现代艺术拼布兴起于欧美，20世纪70年代在欧美流行并逐步产业化，机缝多于手工。中国现代拼布艺术产业近十几年刚刚兴起，多用手缝，如今已上升到创意设计、文化传承的拼布艺术层面。

金媛善老师是中国流行色协会拼布艺术与色彩专业委员会副主任，首席专家。2011年，在金媛善和朱意华等专家的积极倡导下，作为中国唯一的拼布色彩与艺术专业组织成立了，旨在推动拼布传统文化保护和国际民间交流，推动中国拼布艺术的创新、发展、传承，促进艺术和产业结合，进而美化人民生活。拼布专业委员会自2015年以来多次在北京和深圳举办国际拼布与手工艺术展及拼布邀请赛，为广大拼布及手工艺术爱好者和从业者搭建一个展示、传播、学习、交易和分享的国内外交流平台。

始于颜值，敬于智慧，合于性格，久于善良，终于人品，羡于技艺，用此形容金老师和她的作品再贴切不过了。金老师端庄秀丽，气质优雅，智慧善良，性格坚毅，爱憎分明，执着坚韧，热爱中国传统文化，在她身上总能找到各式各样的拼布服饰，包括：戒指、胸花、手包、围巾、服装等，每次公共活动，都会成为焦点人物，吸引无数眼球。而她的作品更令人震撼，惊诧用色和色彩搭配的唯美构图，为精湛的针法而折服，是美的感受，是工匠精神的体现。金老师是中国拼布艺术的代表人物，也是中国拼布艺术和国际交流的一张名片。

金老师是朝鲜族，从小受家庭文化的熏陶，擅长手工，喜爱制作拼布。20世纪90年代开始遍访各少数民族地区，考察、学习各民族传统手工技艺，并将针法、图案和技艺应用在拼布创作中，执着于朝鲜族拼布艺术的研究和创作，并积极推动其他少数民族的拼布艺术研究和发展。金老师拼布采用丝绸、苎麻、亚麻面料，其中丝绸面料全部来自于丝绸服装裁剪后剩下的布块。创作作品时，创意构思、寻找不同布块、按色系分类、不断调整图案，三思而后行，一经确定，坚定执着开始缝制，历经数日甚至数年完成作品。迄今为止，共创作了500余幅拼布艺术品和拼布生活作品。2米以上壁挂作品30余幅，是中国拼布艺术代表人物，作品被不同国家的艺术馆收藏。金老师最得意的拼布作品之一《百花争艳》，获得中国流行色协会"色彩中国2017"中国传统工艺色彩大奖（拼布类），共制作了349朵太阳花，其中，大花256朵，每朵使用了280多块布块；小花90朵，每朵使用40多块布块，采用叠花、平针、倒三针、撩针技艺，将花朵连接成一幅绚丽的百花争艳。这幅作品共计用了7万多块丝绸布块，金老师通过执着与坚持，历时七年，创作出了一幅令人震撼的作品。"百花争艳"构思巧妙，技艺精湛，寓意着各美其美，美美与共，大团圆的美好祝福。金老师最得意的作品还有她送给孙子、孙女的拼布被，一针一线倾注着满满的爱、期望和祝福。

拼布有多种针法，金老师将倒三针技法应用于拼布技艺中，她常说，退一针、再退一针、再退一针，再大踏步前进一步，退一步海阔天空，退三针是为了大踏步前进一步，体现了三思而后行的理念，是传统中华民族智慧和品格的体现，倒三针的技法，是线的完美表达。钱形纹也称"如意纹"，是用方形的布块经过缝合、翻折后形成的铜钱相连的纹样。它综合运用了倒三针、平针的工艺，在"钱眼"处还点缀着打籽绣的花卉，铜钱相接处缀有立体的"小蝙蝠"，"蝠"在眼前，寓意幸福吉祥。

金老师是中国拼布的代表人物，积极参与国际交流，将中国的拼布艺术带到国外展览展示，作品先后被韩国、日本、澳大利亚、美国等国家艺术机构和个人收藏。拼布源于生活，传递故事，在不断发展中寻求美、创造美。拼布艺术，散发着独特的魅力。奇妙构思、创意设计、色彩搭配、寻找面料、缝制拼布的专注和完成的喜悦，每一幅作品都形成独一无二的个人风格，成为一件件极具观赏和审美价值的艺术品。

为了拼布艺术的传承，金老师开始教授学生，她认为这是一种责任。首先是教授做人，她认为作品如人品，要有颗善良的心，传承中国传统文化的责任心，先有德，而后有美；并将积累多年的技艺、针法和图案设计方法传授给学生，鼓励学生开展少数民族图案研究并应用在拼布作品中。金老师多次在少年宫、中小学校和展会上教授中小学生制作钱形纹车挂、太阳花胸花、南瓜花头饰、南瓜花胸针、拼布笔带和拼布书皮等，让孩子们了解中国传统拼布，传播拼布的美丽，培养孩子坚毅和执着的性格。

《母亲的香气》一书，共收集了金老师144幅作品，展示了她精湛的技艺、巧妙的构思、和谐的用色和色彩搭配，是对母亲的回忆，努力、奋斗、自强，也是一代一代文化、传统、技艺创新的传承。它记载了金老师创作的作品寓意和创作感受，无声地讲述着其中的故事，传递着美好情感和心中的梦想，表达着对生活的爱，对自我、对家庭、对社会、对环境甚至对国家以及世界的爱，对美好生活的向往。透过一幅幅作品，仿佛看见她一针一线专注地制作拼布作品，散发出光彩和魅力以及母亲的香气。拼布是一种创意，是一种精神，是一种文化，是一种传播，是一种传承，是一种艺术，是一种美好生活。

希望透过本书，让更多人了解拼布，走进拼布世界。让中国拼布艺术传承、创新和发展。

感谢金媛善老师对中国流行色协会拼布色彩与艺术研究专业委员会工作的支持，也感谢她对中国拼布事业发展做出的杰出贡献。

中国流行色协会 会长
2019年夏

Prologue 2

Patchwork is an art of sewing together pieces of fabric into a completed work with different purposes and connotations. As the needle moves among different shapes of fabrics, it expresses the beauty of colors and patterns. Patchwork is a kind of needlework. Among 56 ethnic groups in China, 33 ethnic groups have patchwork techniques with over 28 kinds of manual stitches. The patchwork pattern of ethnic groups in China covers those popular in foreign countries. The traditional patchwork originated in China. A patchwork from the Chinese Tang Dynasty has been collected in the British Museum. The modern "Patchwork Art" originated in Europe and America. It was popular in Europe and America in the 1970s and gradually industrialized, with more machine sewing than manual sewing. China's modern patchwork industry has just emerged in the past decade, with more manual sewing. Modern patchwork has reached to the artistic level of creative design and cultural heritage.

Mrs. Jin Yuanshan is the deputy director and chief expert of the Patchwork Color and Art Committee of the China Fashion & Color Association, the only professional patchwork color and art organization in China. It was established under the active advocacy of experts including Jin Yuanshan and Zhu Yihua in 2011, aiming at promoting the protection of traditional culture and international non-governmental exchanges, the innovation, development and inheritance of Chinese patchwork, and the integration of art and industry, beautifying people's lives. Since 2015, the Patchwork Committee has held many international patchwork exhibitions and patchwork invitational tournament in Beijing and Shenzhen, setting up a platform for patchwork displaying, communicating, learning, trading and sharing at home and abroad for patchwork lovers and practitioners.

Firstly attracted by her appearance, then fallen for her talents, enjoying her agreeable temperament, faithful to her kindness, finally conquered by her virtues, admiring for her skills, it cannot be more appropriate to describe Mrs. Jin and her works. Mrs. Jin is dignified, beautiful, elegant, intelligent and kindhearted, firm in character, clear in love and hatred, perseverant and obsessed with Chinese traditional culture. She always wears a variety of patchworks, including: ring, brooch, handbag, scarf, garment, etc.. She is the focus of attention on each public event, attracting countless eyeballs. Her works are so impressive that people admire the fabulous composition, color matching and exquisite stitches, which represents the beauty and the craftsman spirit. Mrs. Jin is a representative of Chinese patchwork and a business card for Chinese patchwork and international exchange.

Mrs. Jin is Korean minority. Influenced by family culture from an early age, she is good at handwork and loves to make patchwork. In addition to the research and creation of Korean patchwork, Mrs. Jin also actively promote the research and development of patchwork of other ethnic groups. In the 1990s, she visited various ethnic groups, investigating and learning the traditional ethnic handicrafts, applying ethnic stitches, patterns and techniques to the creation of patchwork. Mrs. Jin uses silk, ramie and linen in her patchwork, of which the silk are the leftovers of the silk garments. When creating a work, from creative ideas, looking for different pieces of cloth, to sorting by color, constantly adjusting the pattern, all needs thinking twice before proceeding, once determined, she began to sew. It usually takes several days or even years to complete a patchwork. So far, more than 500 patchworks and more than 30 wall-hangings with over 2 meters have been completed. As a representative of Chinese patchwork, her works are collected by art galleries in different countries. One of the favorite patchwork of Mrs. Jin, *Hundreds Flowers Blossoming* won the "Color China 2017" Chinese Traditional Crafts Color Award (Patchwork category) held by China Fashion & Color Association. A total of 317 sunflowers were made, among which 227 are middle and large-size flowers, each uses more than 280 pieces of cloth; 90 small flowers, each uses more than 40 pieces of cloth. The stacked flowers, plain stitches, reverse three stitches and running stitches are adopted to connect the flowers into a beautiful art of "hundreds flowers blossoming". A total of more than 70,000 pieces of silk were used. It took seven years of hard work every day to create such an amazing piece of art. The "Hundreds flowers blossoming" is ingenious and exquisite, meaning that each flower has its own beauty, making them together is a beautiful reunion. In addition, the quilts for Mrs. Jin's grandson and granddaughter are also her favorite patchworks, which is filled with her love, expectations and blessings.

There are a variety of stitches for patchwork. Mrs. Jin applies the back triple stitch technique to the patchwork. She often said that step back one stitch, one stitch and one stitch, and take a big step forward. Taking steps back means there will be a bigger world in front of you, the back triple stitch is for the sake of further strides. It reflects the idea of thinking twice before proceeding, embodies the wisdom and character of the Chinese nation. The technique of back triple stitch perfectly expresses the thread. The money pattern is also called "Ruyi" pattern. It is a pattern of coins that are formed by square pieces of cloth after stitching and folding. It integrates both the techniques of back triple stitch and plain stitch, with decoration of flowers embroidered with seeds at the "money eye", and three-dimensional "small bats" at the connection. The "bat" appears in front of us, implying happiness and auspiciousness.

Mrs. Jin is a representative of Chinese patchwork. She actively participates in international exchanges and brings Chinese patchwork to foreign exhibitions. Her works have been collected by art institutions and individuals from Korea, Japan, Australia, the United States and other countries. Patchwork originates from life and conveys stories. It is in the constant development that human being seeks beauty and create beauty. The patchwork exudes a unique charm. With unique concept, creative design, color matching, fabric search and the joy of completion, each piece of work forms a unique personal style, becoming a piece of art with great appreciation and aesthetic value.

In order to inherit the art of patchwork, Mrs. Jin began to engage in teaching. She believed that it was her responsibility to teach. First of all, she teaches students how to be a good person. The style of a work shows the character of a man. A person must have a kind heart, a sense of responsibility to inherit Chinese traditional culture, first virtue, then beauty. Secondly, she teaches the skills, stitches and pattern design accumulated for many years to students. Students are encouraged to study ethnic patterns and apply them in patchwork. The first lesson is about the seven methods of making money pattern. Mrs. Jin has taught students in the Children's Palace, primary or secondary schools and exhibitions to make money-pattern car hangings, sun flower brooch, pumpkin flower headwear and brooch, patchwork pen bag and book cover, so that children can understand Chinese traditional patchwork, spread the beauty of patchwork, and cultivate their perseverance and persistence.

The book *Mother's Aroma*, a collection of 144 works by Mrs. Jin, shows her exquisite skills, ingenious concept, harmonious color matching. It is a memory of efforts, struggle, self-improvement of the mother, also an inheritance of culture, tradition and technological innovation from generation to generation. It records the works created by Mrs. Jin, connotation of her works and her creative feelings. It silently tells the stories behind the art, conveying the beautiful emotions and dreams of the heart, the love for life, for the self, the family, the society, the environment, even the country and the world, the yearning for a better life. Through each piece of work, we can see her concentration on her patchwork creating, which exudes brilliance and charm, as well as the mother's aroma. Patchwork is a kind of creativity, a kind of spirit, a kind of culture, a kind of communication, a kind of inheritance, an art and a life.

I hope that this book will make more people better understand the patchwork and enter the patchwork world, inherit, innovate and develop the Chinese patchwork.

I would like to thank Mrs. Jin Yuanshan for her support of the work of the Patchwork Color and Art Committee, China Fashion & Color Association, and her outstanding contribution to the development of patchwork in China.

the China Fashion & Color Association　Zhu Sha
Summer, 2019

序 3

见素抱朴的拼布艺术

世代母女传承的手工技艺，我们可以称其为"母亲的艺术"。"慈母手中线，游子身上衣。临行密密缝，意恐迟迟归。谁言寸草心，报得三春晖。"母亲手中的针线，毫无功利意图，有的是满满的温情，充满生命的热忱，大家都要向天下的母亲致敬。

我参加民艺调研出版工作多年，在所接触的林林总总民间艺术中，"女红"最让我感动，"母亲的艺术"也是在我眼里最重要的一种艺术。中国是世界上最古老的农业文明之一，数千年"男耕女织"的社会形态造就了人民衣食的生活基础。人们说：妇女撑起半边天。包括纺织在内的女红，对辉煌的华夏文明起了默默地推动作用。"半边天"，并非过誉。

又有人说：现在成衣满街都是，谁还耐烦做女红呀？事实上，正是因为规格化量产成品的泛滥，手工才格外显出珍贵的人情之美。拼布正是这样一种温暖而又充满人情的手工艺术，妈妈们用一针一线，缝入对子女的爱和期盼。

在金媛善老师的拼布艺术里，恰如其分地体现了这一点。"满堂拼布待月明，一盏清茗酬知音。"我与金媛善老师结识于她的拼布艺术展。金老师数十年如一日执着于拼布创作，初见她的作品，被其匀称细密的针脚、独具匠心的色彩搭配所折服。相识之后，我们一见如故，在北京服装学院天工传习馆她的拼布坊中有多次交流，成为知己，时有往来。"作品如人品，先有德后有美"，了解了金老师创作拼布背后的故事，才明白这些美好的作品不只来自于技术，更多的还来自于内心深处的爱，来自于笃定执着的坚持，来自于尽善尽美的要求。

正如金老师曾经给孙子所做的一床被子，在给孙子的信中她写道："这个被子是藏在五彩线里的奶奶的梦想和祝福，两边各有24个小布块，分别表示奶奶今天的24小时和明天的24小时……绗缝的每一针，是希望你在成长过程中学会'忍耐'；坚持一针针绗缝成直线，是希望你将来成为正直的人；一块块将布拼在一起是希望你今后与姊妹、邻居、朋友紧密团结，同时希望你在每一次使用这床被子的时候想起奶奶、父母给予的温暖，希望你在得到爱的同时也学会付出爱心与孝心。"把对儿孙的祝福、祈盼体现在一针一线之中。

又如她的作品《百花争艳》，美丽的花朵色彩绚烂绽放，来自于除旧迎新时的瞬间感怀——"大年三十的交子时分，当看到万紫千红的焰火在空中竞相开放，一年辛勤的劳作都化作满天艳丽的花朵，让人不由得感谢生活的美好，并期盼来年的丰收景象。"针随心走，有了熟练精巧的手艺，再有超乎寻常的艺术追求，如此才能达到一流的境界。

这些拼布之美，也源自于千百年来的深厚积淀。拼布是一种见素抱朴的女孩艺术，是结合生活实际需求的再利用。人们出于节俭，在易破损的位置补缀布片重复使用，平时剪裁中剩下的碎布，妇女们也不会随手丢弃，积少成多，将一块块布头拼接起来，面料和形式都没有限制，再制成衣物以及被面、门帘、包袱等各种生活用品，对物质匮乏的民间百姓来说是一种巧妙的方法。这也契合佛家宣扬的节俭理念，僧人所用袈裟正是一种拼布装，用善男信女捐赠收集的布头缝成僧衣，便演变为"袈裟"。

不仅有物尽其用的好处，打破了以整幅布料制衣理念的拼布，还形成了一种独特、别样的美感，也被越来越多人喜爱。不同颜色布块裁剪成方形、三角形、菱形、圆形，经过细心的拼接设计，焕发了新的生命活力。唐代王维诗中有"裁衣学水田"，世俗妇女也喜爱这种以各色零碎锦料拼合缝制成的"水田衣"，整件服装面料色彩互相交错形如水田，成为时兴的服装样式。

长久以来，拼布融入中华民间习俗生活。比如很多民族都有给小孩制作"百家衣"的习俗。家里有小孩出生或满月，妈妈把亲朋好友送来的一块块碎布拼起来做成孩子穿的衣服或被面，成为"百家衣""百纳被"。大家在送布的时候说上吉祥话，希望孩子穿着大家的衣服，沾得各家的福禄寿喜气，保佑孩子平安百岁。

在飞速发展、物质过剩的今天，拼布依然有着不可或缺的意义。今日"用即弃"成为消费的普遍观念，纺织品服装是增长很快的家庭垃圾，而在纺织服装生产行业中，也产生大量的余料、废料，是否可以节俭拼布的理念，分类再设计加工利用，"变废为宝"？

更重要的是，如果大家能够亲自动手参与碎布的拼接，做成一件件可用可赏、还可以传家的作品，这个过程本身就是一个难得的体验。躲开喧嚣的生活，静下心来拿起针线，感受生活最纯朴的美感，是舒缓压力、寄托情感的休闲，也是一种修行的过程，慢慢地把内心深处的爱挖掘出来，重拾美好。

祝贺金媛善老师的拼布艺术得以结集出版，也希望这一门传统艺术能焕发出新的生命力！

黄永松
2019.5.28

Prologue 3

We call the craftsmanship passed down from mother to daughter for generations "Mother's Art". "The thread in the hands of a fond-hearted mother, makes clothes for her wayward boy; carefully she sews and thoroughly she mends, dreading the delays that will keep him late from home. Such kindness of warm sun, can't be repaid by grass?" The needlework in the mother's hand, no any utilitarian intention, is full of warmth and enthusiasm for life. Everyone should pay tribute to all mothers in the world.

I have engaged in the research and publication of folk art for many years. Among all folk arts that I have contacted, needlework touches me the most. In particular, "Mother's art" is a very important art in my eyes. China is one of the countries with oldest agricultural civilization in the world. The social form of "The men plough and the women weave" continued for thousands of years and created the basic necessities for life. People often say that the women can hold up half the sky. The needlework, including weaving, played a silent role in promoting the splendid Chinese civilization. "Half the sky" is not overstated.

Some people doubt: Now that the readymade clothes are full of streets, who is still patient to do needlework? In fact, just because of the overflow of standardized mass production products, the handwork is particularly attractive as it reflects the precious human feelings. The patchwork is such a warm and humane handicraft that through needle and thread, mothers stitch their love and expectations for their children into the works.

This is appropriately reflected in the patchwork of Mrs. Jin Yuanshan,. "All the patchwork in the house is for the bright moon, A cup of warm tea is for a best friend." I got to know Mrs. Jin Yuanshan in her patchwork exhibition. Mrs. Jin has been dedicated to patchwork creation for decades. When seeing her work for the first sight, I was convinced by its well-proportioned stitches and unique color combinations. We felt like old friends at the first meeting and later exchanged ideas many times in her patchwork studio in Tiangong Inheritance Culture Institution of Beijing Institute of Fashion Technology. "Work is like a person's character, virtue comes before beauty". Upon understanding the story behind Mrs. Jin's creation of patchwork, we can understand that these beautiful works are not only from technology, but also from deep love, from perseverance and obsession, and from pursuit of perfection.

Once she made a quilt for her grandson. Just as Mrs. Jin wrote in her letter to her grandson: "This quilt contains grandma's dream and blessing hidden in the colorful thread. There are 24 small pieces of cloth on each side. They represent 24 hours of my grandmother today and 24 hours of tomorrow. Each stitch is to hope that you will learn to be patient in the course of growing up; Each straight line is to hope you will become a person of integrity; Piecing the patches together is to hope you will be closely united with your siblings, neighbors and friends, hope you will remember the warmth from your grandma and parents every time you use this quilt, hope you will learn to love and be filial while receiving love." The blessings and wishes for the children and grandchildren is embodied in each stitch and tread.

Another representative work is *Hundreds Flowers Blooming*, beautiful flowers blossom in brilliant colors. It comes from the sensation of welcoming the new year. "When I saw the colorful fireworks burst in midair at the New Year's Eve, one year's efforts turned into the gorgeous flowers in the sky. people can't help but thank the beauty of life, and pray for the harvest of the coming year." Following your heart, together with the skillful and exquisite craftsmanship, and the extraordinary artistic pursuit, so that the first-class realm is reached.

The beauty of the patchwork also stems from the deep accumulation of thousands of years. Patchwork is a kind of plain and simple needlework. It is a recycle of waste based on the actual needs of life. People for thrifty often patched pieces of cloth on the places easy to be torn when in hard days. The women would not discard any leftover after cutting the cloth. Many a little make a mickle. The leftovers and rags were pieced together, no restrictions on fabric or form, they were re-made into garments, quilts, curtains, bojagi and other daily necessities. It is a clever way for people who are short of materials. This is also in line with the concept of thrift advocated by the Buddhists. The robe of the monk is a form of patchwork. They are made by pieces of clothes donated by Buddhist devotees, which has evolved into a "cassock".

Patchwork not only has the advantage of making the best use of everything, breaking through the concept of making garment with a whole piece of fabric, but also forms a unique, different aesthetic and is favored by more and more people. The patches with different colors are cut into squares, triangles, diamonds and circles. After careful design and stitching, it has a new vitality. In Wang Wei's poems of Tang Dynasty, there was such words that "cutting cloth is like paddy field". Women also like this kind of paddy field garments made of various leftovers and rags. The color of the whole garment is intertwined like a paddy field and becomes a fashionable clothing style.

Patchwork has been integrated into the Chinese folk customs for a long time. For example, many ethnic groups have the custom of making "a hundred families garment" for children. When a child is born or one month old, the mother uses pieces of clothes sent by relatives and friends to make clothes or quilts for children, it is the so called Baijiayi (a hundred families garment) or Bainabei (a hundred families quilt). Everyone will say auspicious words when they deliver the cloth, hope that the children will wear the garment out of the odds and ends of the cloth sent by everyone, get lucky and live happily, have a long life and good health.

In today's rapid development and material surplus world, patchwork still has indispensable significance. Nowadays, "disposable" has become a common concept of consumption. Textiles and clothing are household wastes that are growing rapidly. In the textile and garment production industry, a large amount of surplus and waste materials are produced. Is it possible to utilize the economic concept of patchwork, redesign, reprocess the waste , "turning waste into treasure"?

What's more, if everyone gets involved in the stitching and piecing of the patches, making a piece of patchwork that can be rewarded and can be family-run, this process itself is a rare experience. Avoiding the hustle and bustle of life, calming down and picking up the needlework, feeling the most simple beauty of life. It is a kind of leisure that relieves pressure and sustains emotions. It is also a process of spiritual practice, which can slowly excavates the deep love inside, and regain the beauty.

Congratulations to Jin Yuanshan's publishing of her patchworks. I hope that this traditional art can be renewed with new vitality.

Huang Yongsong
May, 2019

前言

拼布，英文为Patchwork（又称Piecine、Quilting），是一种将裁剪成规则或不规则的布片按照设计者意图进行连接缝合的工艺，也是一种独特的艺术形式和装饰方式。它的生活属性很强，常被用于服饰和家居用品。

拼布的历史悠久，关于其源起也说法不一。据多年考察研究——从少数民族实地寻访到博物馆资料查阅，我认同多数人所持观点：拼布是起源于东方民间的一种手工艺，11～13世纪传入欧洲。

我国的北方地区是拼布的起源地之一，作为传统女红技法，这种兴起于民间的拼缝补缀制作方式不仅是一种独特的工艺形式，更具有其丰富的文化内涵。其中，最富代表性的民间拼布，是为新生儿制作的服装——百家衣：婴儿诞生后不久，产妇的亲友会到乡邻四舍逐户索要五颜六色的小块布条（若得到老年人做寿衣的边角布料最好）用以拼制成衣。这种风俗可能起源于氏族社会文化遗风，人们认为婴儿在众家百姓、特别是长寿老人的赠予福泽下可以健康成长。

在经济与科技飞速发展的今天，曾经美好而温馨的女红花朵日渐凋谢。也正因此，当我们拿起祖母和母亲当年留下的老物件时，会格外亲切地感受到一股香气扑面而来——那是女人的香气，是女人用善心、巧手、大爱和智慧创造出来的香气。

自幼跟随祖母、母亲学习针线活儿的我，想要将这种香气延续下去。这也是我出这本书，并逐一写出主要作品背后故事的原因。

虽被世人抬爱，称赞我是"中国拼布第一人"，一个人不可以随便讲自己是艺术家。我和拼布发生关系，是因为这些年做了许多拼布作品并参加了一些活动，人家就说"你是做拼布的人"，抵赖不掉。还有人把我说成是"色彩大师"，但唯一令我有切实感的是在拼布中度过的每分每秒——拼布是我的全部热爱与整个人生。

我的母亲是裁缝，我幼时的玩具就是各种各样的布头；在我人生的艰难岁月，拼布成为我最大的精神支柱……在五彩布头的世界中，我用一针一线作为语言倾诉全部的喜怒哀乐和人生感悟。是拼布，让我放慢脚步、平心静气——拼布创作对我而言就像是呼吸一样自然，它是我的最爱、我的生活、我的全部。

在我看来，拼布是一种社会活动、一种文化传承、一种执着的精神、一种艺术创作，同时它也提高了我的动脑、动眼和审美三种能力。这种针与线在五彩布里传情的艺术，令我重拾起对传统文化、古法技艺的信仰。

此书收录了我大部分拼布作品：最早的作品制作于20世纪70年代——我的大儿子出生之前，最新的作品去年才刚刚完成。这其中，有为参加展览而创作的艺术品，也有早年为了养家糊口而制作的工艺品，更多的是为家庭和儿孙们制作的日常生活用品。每一件作品背后都有一个我自己的故事，它们凝聚了我多年的心血与情感，传递着我对生活的热爱与对拼布艺术之美的执着追求。

通过这本书的写作和出版，我想和拼布爱好者以及从事艺术创作的年轻人进行一次心与心的交流：做好一件事的背后，是一生的坚持不懈，是专注求真，是责任与热爱，也是自我的不断超越。技艺要经过多年的磨砺才能成"精"，而感情也在不断地练习和陪伴中愈发真挚深沉。尤其拼布艺术，是美的艺术、爱的艺术、技的艺术，没有捷径可走，只有不断地学习、锲而不舍地努力，才能逐渐接近理想的彼岸。

感激这次成书的过程，令我能借此机会对自己的作品进行一次分类与整理，并重新梳理自己的创作方法与理念，更重要的是，让我又回忆起人生的很多往事，它们是最珍贵的……

2018年夏

Preface

Pinbu, in English it is called Patchwork (also known as Piecine, Quilting). It is a technique to stitch the pieces of regular or irregular cloth together according to the purposes of the designers. It is a unique art form as well as a decoration method. It can be applied to our daily lives. It is usually used as dress articles and household supplies.

Patchwork has a long historical standing. There are also different versions about the origin of it. According to the investigations and researches in the past years — from field visits for minorities to the information query in the museum, I agree with the view of the majority: Patchwork originated from an oriental folk handicraft art. It was then introduced into Europe in the 11~13th century.

The northern part of our country was also one of the areas of origin of patchwork. As the traditional needlework technique, arising from the folk, patchwork was not only a unique technique, but also a production method with rich cultural connotation. Thereinto, the most representative folk patchwork, was the garment made for the newborns — patchwork clothes. Soon after the babies were born, the relatives of the mother would visit the neighbors to ask for small pieces of cloth in different colors (if they could get the leftover cloth of the shrouds of the old, it would be the best) to make garments. This custom probably originated from the social and cultural relics of the clan society. People believed that the newborns could grow healthily under the blessings of many people, especially the longevous.

Today, with the rapid development in economy as well as technology, the once beautiful and sweet needlework is gradually pining away. That was why every time we took up the old objects left by our grandmothers and mothers, we could kindly feel the sweet smell coming over our faces that was the fragrance of women, which was created by women with benevolence, skill, love and wisdom.

I learnt how to sew from my grandmothers and mothers ever since I was a child. I would love to keep the fragrance going. That was also the reason why I published this book and wrote the stories behind the main works one by one.

Though I am loved by people, and praised as "the First Patchwork Artist". I cannot call myself an artist. The reason why I was related to patchwork was because I participated in some activities with some patchworks I have done in these few years, so people would say: You are the patchwork artist. Don't deny it. Some people suddenly called me as "the Color Master". However, the only real thing I could feel, was the every second and every minute I spent in patchwork. Patchwork is my love, is my entire life.

My mother is a tailor. When I was young, my toys at that time were all kinds of fabric scraps. In the rough years of my life, patchwork was my powerful spiritual pillar... In the colorful world of fabric scraps, I poured out all my emotions and reflections on life stitch by stitch. It is patchwork that makes me slow down and keeps me calm. Patchwork for me, is as natural as breathing. It is my love, my life and my all.

From my point of view, patchwork is a social activity, a cultural heritage, a persistent spirit and an artistic creation. It at least improves my abilities of brain, eyes and aesthetic judgment. This art of needles and threads in colorful cloth, makes me regain the faith in traditional culture and ancient techniques.

This book contains most of my patchworks: my earliest work was made in the 1970s — before my oldest son was born, while the newest work was just finished in the last year. There are artworks created for exhibition, artworks for making a living, but more daily necessities for families and children and grandchildren. Behind each patchwork, there is a story of me. They embody my years of efforts and emotions, and convey my love for life and my persistent pursuit of the beauty in the patchwork art.

By the writing and publication of this book, I would love to share my thoughts with the patchwork lovers and the young people who engage in artistic creation: the key to doing one thing well, is to hold on persistently, to focus on pursuing the truth, to be responsible, to love and to surpass ourselves constantly. It takes years to master your techniques. With constant practices and accompany, your emotion will become more sincere and much deeper. In particular, patchwork, is the art of beauty, love, technology. There is no shortcut. You can only approach the other shore of your dreams gradually by constant studies and persistent efforts.

Thanks to the book publication process, I have the opportunity to classify and arrange my works, and tease out my creative methods and ideas. More importantly, it reminds me of my past events, which are very precious...

Jin Yuanshan
Summer, 2018

目录
Contents

001　文化篇　Culture

一、拼布——女性的艺术　/ 002
Patchwork – Art of Women

二、中国少数民族拼布　/ 004
Patchwork of Ethnic Groups in China

三、朝鲜族拼布文化　/ 007
Korean Patchwork Culture

011　作品篇　Works

一、家居　/ 012
Household Articles

二、服饰　/ 070
Garment

三、壁挂　/ 142
Wall-hanging

四、包袱皮　/ 182
Bojagi

五、习作与片段　/ 244
Exercises and Pieces

255　制作篇　Production

　　一、蝙蝠花的制作　/ 256
　　　　Production of Batflower
　　二、太阳花的制作　/ 258
　　　　Production of Sunflower
　　三、针线盒的故事　/ 262
　　　　Story of A Sewing Box

264　作品索引　Index

265　后记　Epilogue

文化篇
Culture

一、拼布——女性的艺术

Patchwork – Art of Women

二、中国少数民族拼布

Patchwork of Ethnic Groups in China

三、朝鲜族拼布文化

Korean Patchwork Culture

一、拼布——女性的艺术

拼布向来是妇女们喜爱的手工艺活动之一。从前妇女们对衣物剩余的布料眷恋不舍，或是因为生活的节省、拮据而将剩余的布料再拼接应用，制作成各种生活用品。由于拼接的是各种色彩与各式花纹的布块，所以完成的整体拼布作品呈现出了变化万千的图案。此外，还填加柔软的填充物，再用针线将层层布面缝合在一起，除了具有装饰性美感外还提供了保暖的功能，因此拼布本质上是将实用性与艺术性有机地结合在了一起。除了近代有少数的男性艺术家加入这样的拼布创作外，从过去到现在拼布可以说都是妇女们精心制作的成果。当然女性天生的美感能力，以及对布料的敏感性和使用针线的精致技巧都是不能忽视的，因此我们可以从"拼布"这种形式的发展来探究如何在拼布创作上透过针和线来展现出属于女性们的故事。

说起拼布的风格与其背后动人的故事，不得不提到那些放弃事业、全身心投入家庭中的传统家庭妇女。她们对梦想与希望的追求并没有停止过，将每一针一线的缝制当成向梦想走近的每一步。所以，每一个作品都体现了女性简朴生活的美德，以及女性独有的气质与追求。

百衲衣

这是为孙子出生百日所做的"百天衣"，采用了绗缝的工艺，一针一线都是奶奶浓浓的关爱。

Bainayi

This is the "Baitianyi" (garment for hundred days) made for the grandson's hundred days. It uses the technique of quilting. The needles and threads represents all the care and love of Grandma.

一、Patchwork – Art of Women

Patchwork has always been one of the favorite craft activities of women. In the past, women were obsessed with the cloth leftover, the remaining fabrics were stitched and applied to make various daily necessities for economic reason. Since the patching is made of pieces of cloth with various colors and patterns, the finished overall patchwork shows a variety of patterns. In addition, soft fillers are added, and the fabric are stitched together with needlework. In addition to the decorative aesthetics, it also provides warmth. Therefore, the patchwork essentially combines practicality and artistry. Except a few male artists in modern times who have joined such a patchwork creation, the patchwork from the past to the present is elaborately prepared by women. Of course, women's natural aesthetic ability, as well as the sensitivity to fabrics and the delicate sewing skills play an important part which can not be ignored. So from the development of the patchwork, we can explore how the stories of women are reflected through needle and thread in patchwork creation.

Speaking of the style of patchwork and the moving stories behind it, we have to mention the traditional women who gave up their careers and devoted themselves to the family. Their desire and pursuit of dreams and hopes have never stopped. Each stitch is a step closer to their dream. Each stitch is filled with the color of life. The feelings hidden in the stitches and the threads of the colorful fabric bring them an opportunity to review the social culture. The dialogue with the needle and the thread when making the patchwork makes the women slow down and let them listen to the words of needle and thread, and feel the peace and calm. Therefore, each piece of work reflects the virtues of women's simple life, as well as women's unique temperament and pursuit.

"拼布"是女性的艺术，它是那些无法用文字书写的女性的自我，陈述着以女性为中心的历史样貌。如非裔美国人的拼布，妇女们用拼布陈述记录着她们家族的非洲根源以及环境变迁下的生活形态、社会意识与人权观等。兴起于19世纪末的"疯狂拼布"（Crazy Quilts）展现出了美国妇女寻找自由的精神与意识。日本和韩国的拼布则是女性精致文化理念的传达。在传统观念中，人们认为拼布只是女性闲暇时的手工制作。直到1971年美国纽约的惠特尼美国艺术馆史无前例地举办了一项拼布展，展出了19世纪至20世纪初的美国拼布作品，那些作品中令人惊艳的几何构图让大家认识到传统拼布的"现代性"潜能，也似乎预言了稍后出现的美国当代抽象艺术风潮，这项展览巡回展出了四年，重新掀起了人们对拼布的狂热，使拼布晋升于艺术的殿堂。进行拼布制作的女性也成为艺术家。从此拼布这项被视为"女性"的艺术创作门类，获得越来越多的重视。

从20世纪下半叶至今，拼布艺术的材料逐渐多样化。在传统拼布日益受到人们欢迎之际，艺术拼布（Art Quilt）也同时展示出了异于寻常的艺术表现之路。探索其原因会发现在国际间成名的拼布艺术家，很少是原来就从事传统拼布制作的，她们大多来自各种不同领域，如镶嵌玻璃艺术、平面设计、绘画、陶艺、服装设计、广告设计，甚至是法律、新闻、财务管理等。因此她们在拼布的创作中不受传统技法、形式与理念的束缚；在感知布料既有的特性中，得以随心运用最大的自由意念；并且在当代艺术理念的感召中，不断地扩张材质表现的范畴，让技法与表现形式随着思维、理念与情感相互融汇串联。

如同书画家们运用笔、墨、纸一样，女性们用针和五彩的布、线以及各种材料的元素，经过艺术性的反复构图、精巧的层次排列技法等进行创作，体现了女性创作者独有的组织心思与精巧的艺术表达。埋藏在其中的是女性的自由思维与情感，包含了她心里的梦想，陈述着她们对自我、家庭、社会、环境，甚至是对国家与世界的关注。所以这些女性们所诉说的故事与历史，希望大家认真地去解读，因为大家的理解一定会给社会带来新的意义。

Patchwork is the art of women. It is the self of women who can't write in words, and states the history of women. For example, African Americans use patchwork to record the African roots of their families and life, social awareness and human rights under changes. The Crazy Quilts, which emerged at the end of the nineteenth century, showed the spirit and consciousness of American women seeking freedom. The patchwork of Japan and South Korea conveys the delicate cultural concept of women. In the traditional concept, people regarded patchwork as just a handcraft in women's leisure time. Until 1971, the Whitney Museum of American Art in New York, USA held an unprecedented exhibition, exhibiting American patchwork from the 19th to the early 20th centuries, the amazing geometric composition of those works made people recognized the "modernity" of the traditional patchwork. And it seems to predict the trend of American contemporary abstract art appeared later. This show, which toured for four years, revived people's enthusiasm for patchwork and elevated it to the pantheon of arts. Women who make patchwork become artists. Since then, patchwork, which is regarded as a category of "female" art, has gained more and more attention.

From the second half of the twentieth century to the present, the materials of patchwork are diversified. As the traditional patchwork has become increasingly popular, Art Quilt demonstrated an extraordinary artistic expression. To explore the reasons, it can be found that the internationally famous patchwork artists rarely do traditional patchwork, most of them come from various fields, such as inlaid glass art, graphic design, painting, pottery, fashion design, advertising design and even law, news, medicine, financial management, etc. Therefore, they are not bound by traditional techniques, forms and ideas in the creation of patchwork. They study the fabric features and use their free will to create. In addition, in the inspiration of contemporary art, the scope of material is constantly expanded, which allows techniques and expressions to be integrated with thoughts, ideas, and emotions.

Just as the calligraphers and painters use pen and ink, women use needles and multicolored fabrics, multicolored threads with various materials, to create patchwork through artistic repetitive composition and elaborate hierarchical techniques. It reflects their unique exquisite mind and artistic expression of female creators. The free thinking and emotions of women devoted to the works embodies the dreams in their heart and their concern about themselves, the family, the society, the environment, and even the country and the world. I hope that everyone will seriously interpret the stories and histories told by these women, because everyone's understanding will bring new meanings to the works.

二、中国少数民族拼布

我从1992年开始赴西南和西北的少数民族地区考察，发现在我国56个民族中，包括汉族在内有33个民族擅于运用拼布技法制作服饰及日用品。更幸运的是发现了多达28种针法。中国民族拼布几乎囊括了世界上现存的绝大多数的拼布方法，尤其是少数民族地区的妇女，至今还在运用这些拼布技法并传承着。可以说，中国的拼布爱好者是非常幸运的，因为我们守着一个巨大的拼布艺术宝库。

从拼布的技法来看，少数民族拼布可分为三大类。

一是拼缝的方法。就是在同一平面上将不同织物连接在一起。其中常见的有长条的连接、方形的连接、三角形的连接以及菱形的连接。藏族、土族和朝鲜族的衣袖，用不同颜色的长条拼成，有"彩虹衣"的美誉。同时，拼缝还可与刺绣相结合，带来更加富有层次的艺术效果。

二是贴布的方法。即将一块织物覆于另一块织物之上然后缝制固定。有折边贴布，也有不折边贴布。不折边贴布的处理方法也是精彩纷呈。按照图案的阴阳，有正向贴布，即贴花；也有反向贴布，即挖花。苗族、彝族、壮族的贴布都非常精彩，各有特色。苗族叫贴布绣，汉族叫补花绣，苏绣中的贴续绣也属这一类。

二、Patchwork of Ethnic Groups in China

Since 1992, I have been visiting ethnic minority areas in the southwest and northwest China. 33 of the 56 ethnic groups, including the Han nationality, are goo-g patchwork to make clothing and daily necessities. Fortunately, up to 28 kinds of stitches were discovered. The Chinese national patchwork almost covers the majority of the existing patchwork methods in the world, especially women in ethnic minority areas are still using and inheriting these patchwork skills. It can be said that Chinese patchwork lovers are very fortunate because we are guarding a huge treasure house of patchwork art.

From the skills of patchwork, ethnic patchwork can be divided into three categories.

The first is the method of patching. It is to connect different fabrics together on the same plane. The most common ones are long connections, square connections, triangular connec-

土族七彩绣女长衫（复制）

近现代青海省互助县
北京服装学院民族服饰博物馆藏
土族妇女一般都穿立领斜襟长袍，长袍两袖由五种不同颜色的布条拼接而成。

Gown with Colorful Embroidery of Tu Minority (reproduced)
Modern, Huzhu County, Qinghai Province
Collection of National Costume Museum of Beijing Institute Of Fashion Technology
Tu women generally wear stand-up collar robes, and the robes are made up of five different colors of cloth.

彝族贴补绣女衣裙

20世纪中期 四川省凉山州昭觉县
北京服装学院民族服饰博物馆藏
凉山彝族女子的服装颜色鲜艳，搭配大胆，对比强烈，而且花纹样式繁多。衣上以羊角、火镰纹贴花为主，风格粗犷又不失细腻。

Female Dress with Patch Embroidery of Yi Minority
mid-20th century, Zhaojue County, Liangshan, Sichuan Province
Collection of National Costume Museum of Beijing Institute Of Fashion Technology
Liangshan Yi minority's women clothing is brightly colored, with bold & contrasting matching, and has a variety of patterns. The clothes are mainly made of horns and fire crepe decals, which is rough and delicate.

tions and diamond connections. The sleeves of Tibetans, Tus, and Koreans are made up of long strips of different colors, which is reputed as "Rainbow Clothing". At the same time, the patchwork are combined with embroidery to bring a more layered artistic effect.

The second is the method of pasting. It is to cover a piece of fabric over another piece of fabric and then sew and fix them. There are folding and non-folding pasting. The treatment of non-folding pasting is also various. According to the yin and yang of the pattern, there is a positive pasting, namely a decal, and a reverse patch, namely brocading. This is called Tiebu embroidery by Miao people while Buhua emboridery by Han people, and Tiexu embroidery in Suzhou embroidery also falls into this category.

三是叠布的方法。即布块折叠成不同的形状拼合或叠加在一起。这是一种神奇的拼布方法，往往可以产生意想不到的效果。其中最令人惊艳的是堆绣和钱形纹。苗族的堆绣也叫"堆花""堆绫绣"，又称"贴绫绣"或"绢绣"。先将用皂角水浆过的绫子剪成小方块，再折叠成三角形，然后把这些小三角一层一层地钉缀、拼缝成各种花鸟图案，装饰在衣领上。这一技法启发了我，经过反复地制作和研究，最终制作完成了我最喜欢的作品之一《百花争艳》。钱形纹是将折叠好的方形布块连接在一起后，再翻折形成一个一个外圆内方的铜钱相连的图案。相同的图案在欧美被称之为"教堂之窗"。在新疆库车博物馆，陈列着一幅出土的唐代盖脸布，也运用了钱形纹，九个一组。钱形纹在韩国叫作如意纹，做法稍有不同。现在欧洲的做法和韩国的一样，中国的钱形纹不仅有很早的出土实物例证，而且在民间广为流传。壮族、彝族、白族、满族均有运用，这是财富的象征。而且制作方法更加丰富，据我在少数民族地区所看到的就已经有七种做法，但最终呈现的纹样都是钱形纹。

The third is the method of folding, folding the pieces of cloth into different shapes to form or stack them together. It is a magical patchwork that often produces unexpected results. The most striking are barbola and money pattern. This is called Tiebu embroidery by Miao people while Buhua emboridery by Han people, and Tiexu embroidery in Suzhou embroidery also falls into this category. It is to cut the silk starched with saponin into small squares, and folding them into triangles. Then these small triangles are stapled and seamed layered by layered into various flower and bird patterns, decorated on the collar. This technique inspired me. After repeated production and research, I finally produced one of my favorite works, *Hundred Flowers Blossoming*. The money pattern is a pattern in which the folded square pieces of cloth are joined together and then turnover and form a copper coin with outer circle and inner square. The pattern is called "the window of the church" in Europe and America. In the Museum of Kuche, Xinjiang, an unearthed face-covering cloth from the Tang Dynasty also uses money pattern with nine of one set. In Korea, the money pattern is called the "Ruyi"(as-you-wish) pattern, which is slightly different in making compared with China. Now, the practice in Europe is the same as that in South Korea. The Chinese money pattern not only has a very early unearthed physical illustration, but is also widely circulated among the people, Zhuang, Yi, Bai minorities, and Manchu all use such pattern, which is a symbol of wealth, and the production methods are more abundant. As I have seen in ethnic minority areas, there are seven practices, but the final pattern is the money pattern.

苗族堆绣鸟纹女上衣（局部）
20世纪中期 贵州省台拱镇
北京服装学院民族服饰博物馆藏
此件上衣最精彩的部分就是领口、肩部和袖口的堆绣。堆绣也叫"堆花""堆绫"，此件上衣堆有鸟和鱼的造型，细致而生动。

Miao Women's Embroidered Birdprint Blouse (part)
Mid-20th century, Taigong Town, Guizhou Province
Collection of National Costume Museum of Beijing Institute Of Fashion Technology
The most exciting part of this blouse is the barbola on neckline, shoulders and cuffs. The barbola is also called "Stacked Flowers" and "Stacking". This piece of top has a shape of birds and fish, which is detailed and vivid.

除了拼布的技法，少数民族拼布运用的针法也非常独具匠心。非常典型的就是"倒三针"的方法，也叫作"退三针"。在中国民间和韩国，倒三针运用在拼布或刺绣作品的边角上，不仅用作装饰还起到加固的作用。在我们少数民族地区，还可以看到倒一针、倒二针、倒四针、倒五针、倒六针（元代出土文物，李雨来先生的藏品），都是为了增加美感和加固而做的。但我认为，倒三针不仅兼具美观和实用，还体现了中国人的民族品格。中国不仅有"退一步海阔天空"，还有"三思而后行"。倒三针是倒一针，倒一针，再倒一针，总共往后退了三针，然后往前走一针，最后这一针比倒的三针距离还大一点。所以说，经过退三步反复思考后再大步迈出去，这就是国人"以退为进""三思而后行"的理念。倒三针，这种民间的传统针法并非拼布的惯用针法，我之所以用这种针法，是希望中国的拼布具有本土化的独特风格，体现我们中华民族经反复思考后稳稳当当地往前走的民族风格。

In addition to the technique of patchwork, the stitching method used by ethnic minorities is also very original. The very typical method is the "reverse three-stitch"method, also known as "back three-stitch". In Chinese folk and South Korea, the three stitches are used on the corners of patchwork or embroidery. It's not only for decoration but also for reinforcement. In our minority areas, we can also see that reverse one stitch, two stitches, four stitches, and five stitches (Unearthed relics of the Yuan Dynasty is from the private collection of Mr. Li Yulai) are all used to increase beauty and reinforcement. However, I believe that the reverse three stitches are not only beautiful and practical, but also reflect the Chinese national character. There is a saying in China that "a step back, you'll get room for your next step!" and "think twice and then go". The reverse three stitches are one stitch back, one stitch back, and one stitch back— a total of three stitches back, and then one stitch forward, the stitch forward is a little larger than the three stitches. Therefore, after three steps of rethinking and then striding out, This is the Chinese philosophy of "Retreat in order to advance" and "Think before you act". The traditional folk stitching method of "back triple stitch" method is not the usual way of patchwork. The reason why I use this method is that I hope Chinese patchwork can show a unique local style, which reflects our nation's style of moving forward steadily after repeated thinking.

白族钱形纹拼布围兜

20世纪中后期 云南省大理白族自治州
北京服装学院民族服饰博物馆藏
这件围兜的下半部分，用白色和彩色的布块拼接成相连的铜钱纹样，称为"钱形纹"或"如意纹"。

Money Pattern Patchwork Bib of Bai Minority
Mid-to-late 20th Century, Bai Autonomous Prefecture, Dali, Yunnan
Collection of National Costume Museum of Beijing Institute Of Fashion Technology
The lower part of the bib is made up of white and colored pieces of cloth, and connected to the pattern of copper coins that is called "Money Pattern" or "Ruyi Pattern".

花苗上衣披肩（局部）

20世纪早中期，贵州省毕节市
北京服装学院民族服饰博物馆藏
这种披肩以毛线织花或蜡染布为底，用挑花、蜡染、镶嵌、编织等手法做成纹样。色调以红、黄、黑为主，每种纹样及组合图案均有特定的文化含义，披肩上的红色条和黄色条分别代表长江、黄河，其中长方形纹样代表城池，显示着苗族人对祖先和历史的纪念。

Flower Shawl of Miao Minority (part)
Early and middle 20th century, Bijie County, Guizhou Province
Collection of National Costume Museum of Beijing Institute Of Fashion Technology
This shawl is made of wool woven or batik cloth, made by means of pointelle, batik, inlay, weaving, etc. The red, yellow and black are the main color. Each pattern and combination pattern has a specific cultural meaning, and the red bars and yellow bars on the shawl represent the Yangtze River and the Yellow River, respectively. The rectangular pattern represents the city, showing the Miao people's commemoration of their ancestors and history.

除了拼布的技法和针法，少数民族拼布的色彩和图案往往和该民族的神话传说、图腾崇拜、宗教信仰有关，所以少数民族的传统拼布除了色彩斑斓、形式多样之外，还渗透出神秘的色彩，也更加耐人寻味。俗话说"外行看花哨，内行看门道"，要真正看懂少数民族拼布的"门道"，还有必要了解与该民族相关的历史与文化。例如土族的妇女爱穿五彩花袖的长袍，从肩部到袖口，用红、绿、黑、黄、白五种颜色的布条或彩缎拼接而成，这些色彩的排列不能更改，红色象征太阳位于正中，两边依次是象征森林和草原的绿色、象征大地的黑色、象征五谷丰登的黄色，袖口是白色的，象征纯洁善良的心和生命之源的清水。又如贵州毕节苗族披肩上镶绣的红、黄色条纹，象征着长江、黄河，其中长方形纹样代表城池，显示着苗族人民对祖先和迁徙历史的缅怀与纪念。

In addition to the techniques and stitch method, the colors and patterns of ethnic patchwork are often related to the myths and legends, totem worship, and religious beliefs of the ethnic group. Therefore, the traditional patchwork of ethnic minorities, in addition to various colors and forms, also reveals intriguing mystery. As the saying goes "The insider knows the ropes, while the outsider just comes along for the ride." To truly understand the "ropes" of ethnic patchwork, it is necessary to understand the history and culture associated with the ethnic group. For example, women of the Tu minority love to wear robes with colorful sleeves. From the shoulders to the cuffs, they are made of five colors of red, green, black, yellow and white. The arrangement of these colors cannot be changed. The red one is the sun in the middle, and the two sides are green which symbolizes the forest and the grassland, the black symbolizes the earth, the yellow symbolizes the grain, and the white cuffs symbolize the pure and kind heart and the clear water which is the source of life. Another example is the red and yellow stripes on the shawl of the Miao minority in Bijie, Guizhou, which symbolizes the Yangtze River and the Yellow River. The rectangular pattern represents the city, showing the Miao people's commemoration of the ancestors and migration history.

中国是多民族的国家，多彩纷呈的民族文化造就了巨大的拼布宝库。这些民族拼布具有显著的民族性和区域性。尤其是民族性，构成了其璀璨夺目、缤纷多彩的鲜明的特色。做出这些作品的每一位前辈都是从漫长的历史发展中，通过自己的劳动和智慧创造出别具特色的手工技艺。这些独特的手工技艺可以帮助我们再创新，让传统的工艺焕发出新的生机。可以想象，当我们的拼布爱好者早日投入到我们国家本土拼布技艺的传承和创新之时，将会幻化出令世界拼布界叫绝的效果。所以欣赏、研究、继承和发扬中国本土拼布文化，就成了一项富有意义的课题。

China is a multi-ethnic country, and the colorful and diverse national culture has created a huge treasure house of patchwork. These national patchwork have significant nationality and regionality. In particular, the national character constitutes its resplendent, colorful and distinctive features. Each predecessor who made these works created a unique craftsmanship through their own labor and wisdom from long-term historical development. These unique craftsmanship can help us innovate and bring new vitality to the traditional crafts. We can imagine that when our patchwork lovers are committed to the inheritance and innovation of our native patchwork, they will cause the amazing world-wide effect. Therefore, appreciating, studying, inheriting and carrying forward the Chinese native patchwork culture has become a meaningful topic.

三、朝鲜族拼布文化

朝鲜族最典型的拼布应用是包袱皮（Bojagi），意为布块拼接的包袱皮。包袱皮是朝鲜族传统的生活用品，其本来用途是为了包东西或盖东西以方便运输或防止灰尘。随着历史的发展，如今已衍生出多种用途，它不仅为生活所用，还反映着朝鲜族的思想、文化、礼仪及信仰。其基本材料是棉布、苎麻、大麻、真丝或纸。式样、图案和色彩均富有浓郁的民族特色和地方风格。

三、Korean Patchwork Culture

The typical patchwork of the Koreans is Bojagi, namely, the wrapper patching with pieces of cloth. Bojagi is a traditional Korean household item, which was original used to pack or cover things for facilitating transportation or preventing dust. With the development of history, it has been developed into a variety of uses, not only used for life, but also reflects the Korean minority's thoughts, culture, etiquette and beliefs. The basic material is cotton, rami, hemp, silk or paper. Patterns and colors are rich in ethnic and local style.

包袱皮起源于朝鲜族过去居住的小空间，人们希望用最简便的生活用具来美化房间。包袱皮使用起来很方便，制作简便且费用低，保管起来也很方便，叠放即可。在民间的叫法很多，如包袱皮、包皮布、褓子、福、褓子衣等。因为"袱"与"福"谐音，所以在民间信仰中它象征着"包福""招福"，是一种吉祥的物品。

Bojagi originated from the Korean people who lived in a small space in the past, who did hope to use the simplest living utensils to decorate the room. Bojagi is easy to use, simple to make, easy to store, stackable and has low cost. There are many names in the folk, such as "Baofupi","Baopibu","Baozi", "Fu","Baoziyi" and so on. Because "Fu" and "Blessing" are homophonic in Chinese, it symbolizes "wrap blessing" and "Blessing" in folk beliefs. It is an auspicious item.

包袱皮的制作方法多数是利用做成衣时剩余的边角料来拼花、绣花、印花等制作而成。先将边角料拼出基本规格，例如，正方形、长方形、三角形、多边形等，再按照使用要求设计出大小不一样的包袱皮。其图案和色彩大部分不会特意设计，一般按制作者的感觉，随意把几百个布头自由地拼缀在一起。虽然没有规律，但给人的感觉并不散漫，而是有着内在形式美规律的装饰图案。其图案一般与自然相结合，有花草鸟虫、蓝天白云等。在图案色彩的使用上以素色为主，但可以看到一年四季的"春、夏、秋、冬"，其组成既有华丽的色彩，也有暗淡的色彩。

Most of Bojagi's production methods use the leftover pieces in the clothes to make flowers, embroiders, printing, and so on. First, the leftover pieces are assembled into basic shape, such as squares, rectangles, triangles, polygons, etc. And then Bojagi's sizes is designed according to the requirements of use. Most of its patterns and colors are not specially designed, generally, hundreds of cloth pieces are freely patched together according to the feeling of the producer. There is no law, but instead of looking sloppy, it is aesthetic with the inner beauty. Its pattern is combined with nature, like flowers and birds, blue sky and white clouds and so on. In the use of design color, give priority to plain colors. But the colors can indicate the "Spring, Summer, Autumn, Winter", combining gorgeous and dull hues.

包袱皮是朝鲜族日常生活中不可或缺的生活用具，除了包裹东西外，做完饮食后也常用包袱皮盖上，防止异物落入。另一方面，包袱皮也体现出整个民族的礼仪、信仰。在婚礼等重要的礼仪场合以及与尊贵的人往来时，使用包袱皮包裹表现了诚意与尊重。

朝鲜族的包袱皮（Bojagi），用于包裹物品，便于携带。旧时人们出行，常用其包裹物品，手提或顶在头上都很方便。
Bojagi in Korean minority of China is used to wrap items for easy carrying. When people traveled in old days, they used Bojagi to wrap things. And it was convenient to be crried in hand or put on head.

Bojagi is an indispensable tool in the daily life of the Korean. In addition to wrapping things, it is also used to cover the food to prevent dusts or other things falling in. On the other hand, Bojagi also reflects the etiquette and belief of the minority. In important ceremonies such as weddings as well as meeting with distinguished people, Bojagi parcels shows sincerity and respect.

包袱皮根据材料和使用场合的不同，呈现出两种鲜明的风格。一种素朴无华，用棉布、苎麻和大麻制作，色彩比较单一，多为民间日常所用。另一种则色彩华丽，更多使用真丝的缎、纱、绫和精细的苎麻，再加上绣花、印花，而且一年四季使用的面料也不尽相同，主要用于重要的礼仪场合。

Bojagi presents two distinct styles depending on the diffevent ematerial and the occasion of use. One is simple and unpretentious, which is made of cotton, ramie and hemp. The color of this style is relatively simple. And it's mostly for the daily use. The other is the gorgeous color, which is made of silk satin, yarn, enamel and fine ramie, and then plus embroidery and printing. what's more, the materials is different as the seasons changing. And it's mainly used in important ceremonial occasions.

从朝鲜族妇女制作出来的包袱皮图案和色彩能看到她们对生活的理解、憧憬和希望。包袱皮是朝鲜族最具代表性的手工艺品，也属于高档商品。由于都是各家用日常搜集的各种面料包袱皮还是朝鲜族婚庆等重要礼仪场合中不可缺少的吉庆用品。朝鲜族传统婚礼中的奠雁礼，即是新郎到女家迎亲，献雁（木雕的大雁）为赞礼，象征新婚夫妇像大雁一样爱情忠贞、永不分离。大雁常用精心制作的包袱皮包裹。

Bojagi is also an indispensable auspicious supplies for important ceremonial occasions such as weddings. The geese ritual in the traditional Korean wedding ceremony is that the bridegroom and the female family greet each other, and the geese (wood eagle geese) are the rituals, symbolizing that the newlyweds are loyalty as geese and never separate. The geese often is packed with elaborate Bojagi.

制作，所以图案各不相同，体现出的情趣品味和美学意境也不尽相同。

The patterns and colors of Bojagi made by Korean women reflect their understanding, expectation and hope for life. Bojagi is the most representative handicraft of the Korean and is also a high-end product. Because they are all kinds of fabrics that are collected by households in daily life. The patterns are different, reflecting different tastes and aesthetics.

在世界拼布领域，朝鲜族的包袱皮不仅是一种传统的手工艺品，也是有着高品位的文化艺术品，常被文人雅士作为相互馈赠的首选之物。在美国、加拿大、日本等国家，朝鲜族的包袱皮均有着很高的声誉。包袱皮的拼布里包含着朝鲜族几千年的文化内涵。从其色彩和图案中可以看出，朝鲜族妇女们卓越的审美力、造型力与创造力，充分体现了她们的创作智慧。

In the world, the Korean Bojagi is not only a traditional handicraft, but also a high-grade cultural works and artworks. It is often chosen by scholars as the first choice of gift. In the United States, Canada, Japan and other countries, the Korean Bojagi has a high reputation. Bojagi's patchwork contains the cultures of the Korean for thousands of years. It can be seen from its colors and patterns that the Korean women have excellent aesthetic, styling ability and creativity, which fully reflect their creative wisdom.

朝鲜族妇女们从包袱皮的造型美中找到了现在和未来生活中兼具的实用性和艺术性，进一步发展了民族的传统手工艺术品，使它焕发出新时代的光辉。虽然传统的民间工艺随着时代的发展已逐渐从现代人的生活中分离出来，大部分已成为纯欣赏性的工艺品，其实用性与时代性的功能正不断被现代工艺设计所代替，不再成为时代消费的主流。但作为民间工艺品的包袱皮依旧在现代生活中占据着重要的位置，其原因在于包袱皮是传统工艺的艺术性和现代生活实用性的完美结合，既满足人们多层次的审美需求，又满足现代人的实用需求。朝鲜族的包袱皮作为一种纤维艺术打破了传统的规范，具有独创性和实用性，是女性智慧的结晶，更是当代时代精神和多种美学的反映。

The Korean women find the common practicality and artistry of Bojagi both in the present and future life, and develop the traditional handicrafts to make it show its value in the new era. Although the traditional crafts have gradually separated from the modern people's life as the time goes, most of them are just for pure appreciation, their practicality and the function in the old times are constantly being replaced by modern craftsmanship, and it is no longer the mainstream of consumption in the era. However, Bojagi, a folk crafts, still occupies an important position in modern life. The reason is that the artistic combination of Bojagi's traditional craftsmanship and modern practicality meet the multi-level aesthetic and practical needs of modern people. The Korean Bojagi breaks the traditional norms, is original and practical, and is the crystallization of women's wisdom. It is the reflection of the contemporary spirit of the times and a variety of aesthetics, and also represents a kind of fiber art.

作品篇
Works

一、家居
Household Articles

二、服饰
Garment

三、壁挂
Wall-hanging

四、包袱皮
Bojagi

五、习作与片段
Exercises and Pieces

一、家居
Household Articles

拼布已经成为我生活中不可或缺的组成部分。在日常的起居中，处处可以看到拼布的影子。尤其是盖的被子和挂的窗帘，体量较大，更是拼布创作的重要载体。民间一直有为孩子做"百家被"的传统，所以我耗时很久，为儿子、孙子、孙女制作了拼布的被子，不仅仅希望他们用以取暖，更希望他们从中学到为人处世的道理，感受到来自母亲、奶奶深深的祝福。

　　Patchwork has become an integral part of my life. In the daily life, you can see the patchwork everywhere. In particular, the cover quilt and the curtains have a large volume and are an important carrier for patchwork. The folks have the tradition of making "Baijiabei" for children. So I spent a long time making quilts for my sons, grandsons and granddaughters, not only hoping that they will be warm, but also hope that they will learn from others. Feel the deep blessings from mother and grandmother.

姹紫嫣红　被面

绡，200cm×190cm，倒三针、撩针，1999 年
2007 年日本东京国际拼布博览会传统部门二等奖

  这是给孙女制作的被子，每天完成一部分，耗时一年时间。在给孙女的亲笔信中，奶奶表达了自己的心愿：希望小孙女长得像鲜花一样漂亮、像仙女一样美丽，在温暖的小家庭中起到凝聚、团结的作用，拥有一个让人羡慕的温暖的家庭。

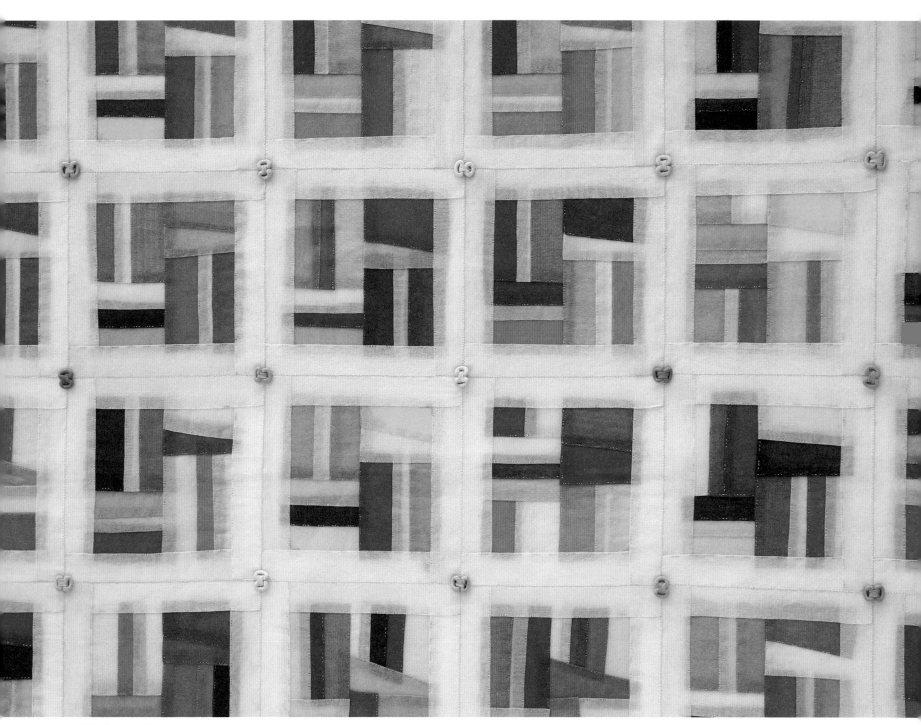

(局部 Details)

Colorful World, Quilt cover

Raw silk, 200cm×190cm, Back triple stitch & Slip stitch, 1999

Second prize in the traditional part of Tokyo International Patchwork Expo 2007

  This is the quilt for my granddaughter, which took me one year to complete. In the letter to my granddaughter, I expressed my wish: I hope that the little granddaughter looks as beautiful as a flower, as beautiful as a fairy, playing a cohesive and united role in a warm family, and has an enviable warm family.

(局部 Details)

017

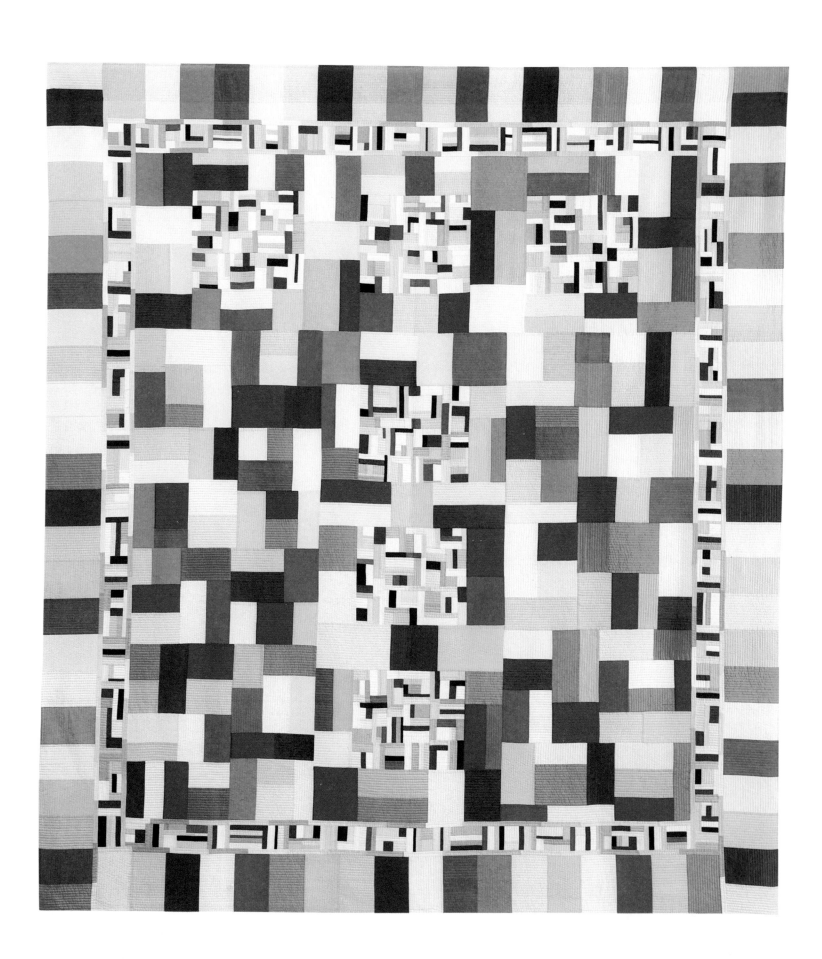

母亲的香气―金媛善拼布艺术

018

虹 棉被

绸、棉花，230cm×200cm，绗缝，1998 年
2007年日本东京国际拼布博览会拼布入围奖

　　我为孙子制作的棉被，耗时五年完成。这个被子的原名是"五彩线里的梦想"，孙子三岁时，他自己将其改名为彩虹的"虹"。

　　我在制作这件作品的同时，还写给孙子一封信，原文如下：

　　藏在五彩线里的奶奶的梦想和祝福送给你可爱的孙子！！！
　　奶奶送给你什么礼物呢？？
　　送给你奶奶一生想要努力实现的梦和爱。
　　奶奶是用五彩的线和五彩的布一针一线亲手做的绗缝拼布彩虹被，用了五年的业余时间用一针一线用心去制作的被。
　　这一针一线里有奶奶自己的梦想，
　　这一针一线里有奶奶的爱心，
　　这一针一线里有奶奶特有的香气。
　　可爱的孙子，你一定闻得到奶奶的香气！奶奶的心永远是爱你的。
　　祝福你一生富贵平安……
　　可爱的孙子，你一定要记牢奶奶制作彩虹被时的五条心愿。
　　第一条，绗缝被中使用的同样的、反复的每一针针法是希望你在成长的过程中学会"忍耐""坚持"。福气在忍耐里。只有坚持和努力，再加上毅力才能成功。
　　第二条，绗缝被中一针一线缝制的、没有误差的每一行直线是希望你将来成为"正直"的人。做人处事一定要学会不考虑计较前后，只要自己对自己负责并保持正直、诚实。坚持如此，奶奶认为你会是人生路上走在最前面的人。
　　第三条，绗缝被中一块一块五彩的布被拼在一起，是希望今后你与姊妹、邻居、朋友之间要紧密"团结"。五彩的布拼在一起后的美丽和紧密团结后的力量……给人们心灵上的感受是同样的。
　　第四条，希望你学会宽容别人，不要只满足于别人对你的宽容。一定要学会自己能原谅自己，但同时也要严格要求自己。
　　第五条，希望你每一次使用这温暖的彩虹被时都会感受到奶奶和父母给予你的温暖和爱。同时希望你得到爱的时候也一定要学会付出爱心与孝心。
　　下面奶奶把制作彩虹被时的真实的想法告诉你。
　　（最上面两侧的）这两个四方块是代表你的爸爸妈妈看着你成长。
　　这中间四个小布头拼成的四方块是代表你一年四季成长的过程。
　　（外围的）这个大四方块是代表奶奶一生的春、夏、秋、冬。在这里有奶奶一生的梦想和爱，也有奶奶日常生活中的疲劳和痛苦，以及奶奶一生藏在自己心里的未完成的梦想。
　　外边左右两侧的24块拼布是奶奶今天的24个小时和明天的24个小时。明天的24个小时任何人也不知道会发生什么，所以明天就是代表梦想。
　　上、下18块拼布是代表你的一代及下一代都要生活富裕（发财）的意思。
　　可爱的孙子，当你长大到18岁时，能考上中国清华大学或美国哈佛大学时，你可以打开翻一翻奶奶一生的日记，在奶奶的日记中你会看到奶奶一生的喜怒哀乐和奶奶一生要求自己做人的原则……
　　奶奶希望孙子刚强！勇敢！一定要成为正直的人！
　　最最爱孙子的奶奶

金媛善

一九九八年七月七日

(棉被背面　Back of quit)

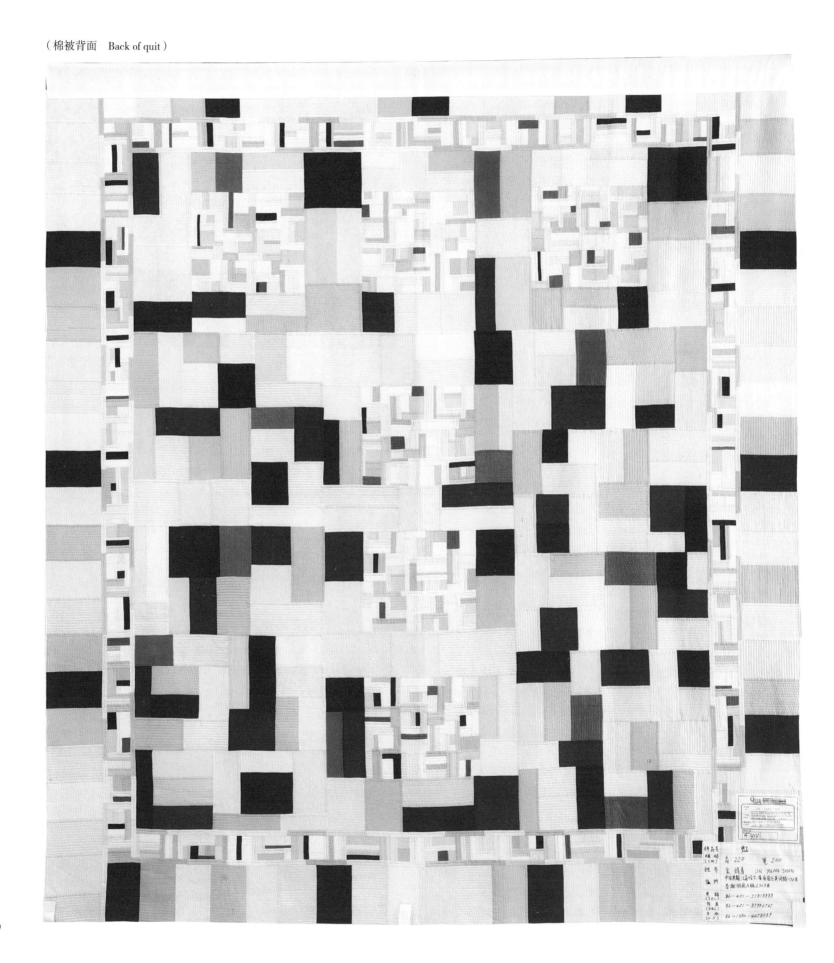

母亲的香气—金媛善拼布艺术

Rainbow, Quilt

Silk & Cotton padding, 230cm×200cm,
Quilting, 1998
Finalist Award of 2007 Tokyo International Patchwork Expo

It took five years to complete the quilt for my grandson. The original name of this quilt was "the dream in the colorful threads." When the grandson was three years old, the name was changed to "rainbow" by grandson himself. At the same time, I wrote a letter to my grandson:

Grandma's dreams and wishes hidden in the colorful threads are given to you, my lovely grandson！！！
What gift does grandma give you？？
Give the dreams and love that your grandmother wants to achieve and obtain in lifetime.
This is a patchwork named rainbow and made with multicolored threads and multicolored fabrics by grandma. Grandma spent five years in spare time to make the quilt.
There is the dream of grandma.
There is the love of grandma.
There is a unique aroma of grandma.
Cute grandson, you must smell the aroma of your grandmother left in the quilt! Grandma's heart will always love you.
Wishing you a wealth and peace life...
My cute grandson, you must remember the five wishes of Grandma to make this rainbow quilt:
Firstly, the same stitch is repeatedly used in the quilt. It is hoped that you will learn to "endure" and "persist" in the process of growing up. Blessing is in endurance. Perseverance, effort and willpower make success.
Secondly, Each line of quilting without error hopes you become a "righteous" person. You must learn to be a person doing things without considering too much situation, as long as you are responsible for your integrity and honesty. Then you will be the first mover.
Thirdly, The pieces of colorful cloth being put together in the quilt means my hope that you will be "solidarized" with your sisters, neighbors and friends in the future. The multicolored cloths and the power of unity have the same meaning.
Forthly, I hope you learn to be tolerant of others. Don't just be content with the tolerance of others. You must learn to forgive yourself and also be strict with yourself.
Fifthly, I hope that every time you use this warm rainbow quilt, you will feel the love from grandma and parents. I hope that when you get love, you must also learn to give love and filial piety.
The real thoughts of grandma when making the rainbow quilt:
The two squares on the top (the top two sides) represent your mom and dad looking at your growth.
The four little squares in the middle represent your growth all year round.
The (outer) big four square is the spring, summer, autumn and winter that represent grandma's life. There is the dream and love of grandma's life, fatigue and pain in grandmother's life, as well as unfinished dreams that grandma has hidden in heart for a lifetime.
The outer 24 pieces of patchwork are 24 hours for today and 24 hours for tomorrow. No one knows what's going on in 24 hours tomorrow, so tomorrow is a dream.
The upper and lower 18 pieces expects you and your next generation can make a fortune.
My grandson, when you are 18 years old and go to Tsinghua University in China or Harvard University in the United States, you should open the diary of your grandma. Where you will see the joys and sorrows of grandma's life and the principle of being a man...
Grandma wants her grandson to be strong! Brave! Be a person of integrity!
Your grandma

Jin Yuanshan
July 7, 1998

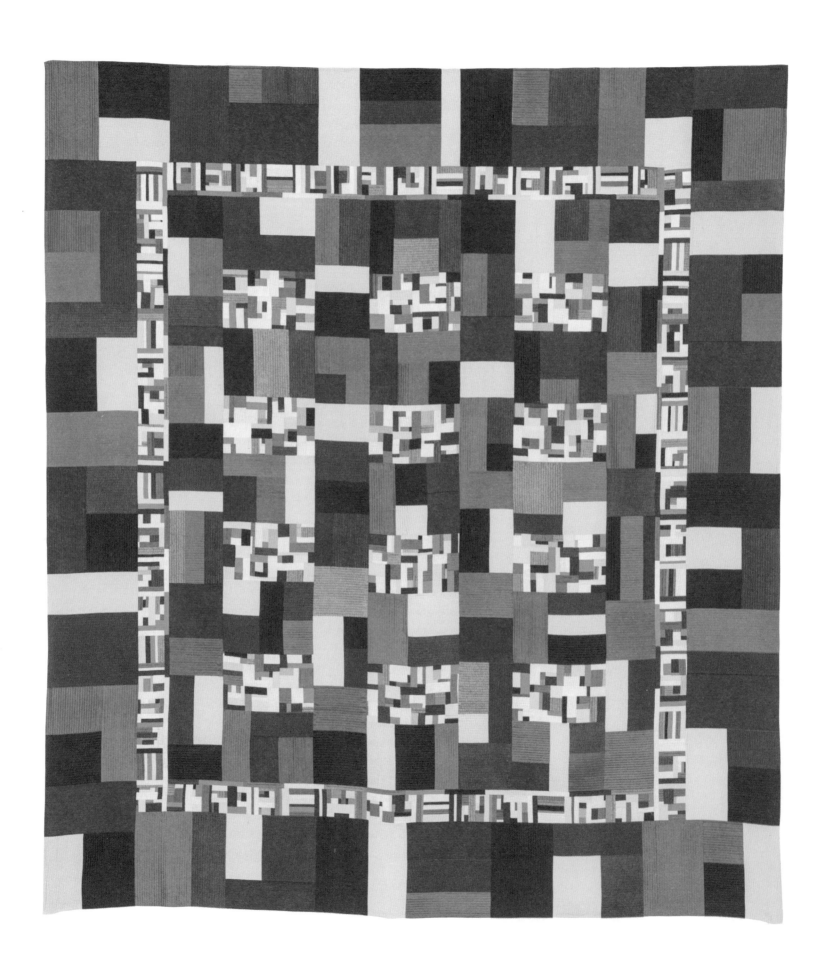

母亲的香气―金媛善拼布艺术

琴瑟和鸣　棉被

绸、棉花，230cm×200cm，绗缝，1978年

　　《琴瑟和鸣》这个名字，是丈夫起的。1973年，我的丈夫被确诊为眼底癌。最初的那段时间，我们每天都在灰暗的病房中，承受着压力、焦虑、苦闷和无奈。我决定做一床拼布绗缝被子，希望他有生之年，可以盖上这床我亲手为他制作的被子，不离不弃、相谐相合地生活下去。

　　之后，每晚我们夫妻在台灯前对坐，我做拼布、他讲故事，两个人说说笑笑，仿佛有讲不完的话。他说："你的针线，走在布上淋漓至极，有阳光的烂漫、有色彩的华美，有生活的真情。"

　　做拼布对我而言是一种将养和解脱，一针一线地缝着时，我心中的痛苦、烦恼都会消失，取而代之的，是一种宁静的舒服、寂寞的享受……

　　我将自己的身心投入到拼布中，作品如实地反映了我的心情思绪；而拼布，也给我了阳光和希望——这张《琴瑟和鸣》被，每年过年时拿出来，夫妻两个人同盖，一共盖了19年。

　　我想，或许是当时做这床被子时我真心的祝福和希望在冥冥中感动了上苍，令一位癌症患者奇迹般地又同我一起共度了20载人间寒暑。

　　我由衷地感激拼布。不是我选择了它，而是它选择了我。

Conjugal Bliss, Quilt

Silk & Cotton padding, 230cm×200cm, Quilting, 1978

  The work *Conjugal Bliss* is named by my husband. In 1973, my husband was diagnosed with fundus cancer. During the early days, we were stressed, anxious, depressed and helpless in the dark ward every day. I decided to make a bed of patchwork-quilt for him. I hope that he can use this quilt and we can live a harmonious life with each other.

  After that, every night, my husband and I sit in front of the lamp. I do patchwork, he tells the story. Two people talk and laugh. There are so many things we want to share. He said that, "Your needlework, threading the cloth to the extreme, like the sun is shining, and the color is gorgeous, and it has the true feelings of life".

  Doing patchwork is a kind of liberation and relief for me. When I do my work stitch after stitch, the pain and trouble in my heart will disappear. Instead, it is a kind of quiet, comfortable and lonely enjoyment.

  I devote myself into patchwork, the work can reflect my thought authentically. Patchwork gives me sunshine and hope. This quilt, I took it out every year; we have used it together for 19 years.

  I think, perhaps it was my heartfelt blessings and hopes when I was doing this quilt. I think I moved the God, and he permits a cancer patient miraculously to spend another 20 years with me.

  I am sincerely grateful for patchwork. I did not select patchwork, but I was selected by it.

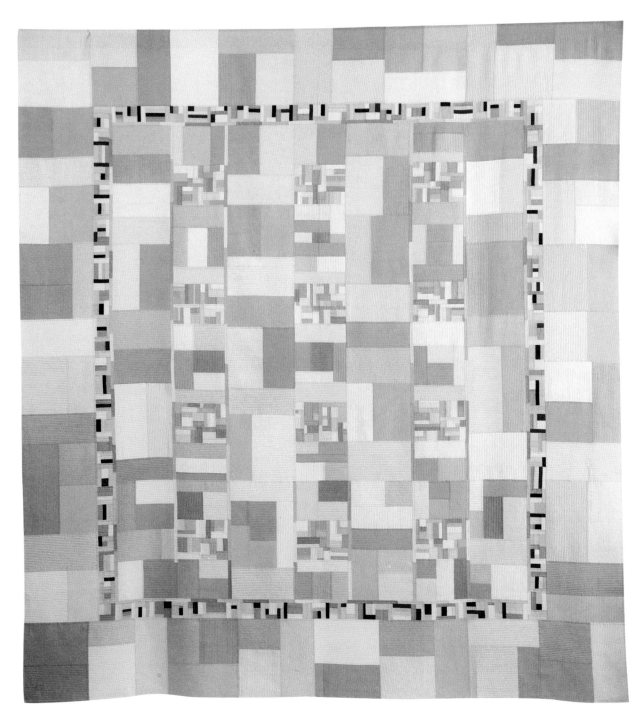

和谐 I　棉被

绸，253cm×236cm，绗缝，1990 年

　　给大儿子和儿媳的被，耗时三年完成。我将自己的一生都藏在了这五彩的布和线里，用五彩的线讲述自己琐碎细密而充满爱意的情思，这五彩的线装点了我的人生。在这彩线和彩布连接拼缝而成的棉被里，是妈妈倾注的暖暖的爱，希望你们相亲相爱，希望你们健康幸福。祝福小两口在美满幸福的新家庭里永远相互尊重、信任、帮助，永远像拼布连接在一起一样相亲相爱。同时，希望以后家庭的每一个成员都能与大自然和谐相处。我把最美好的心愿用一针一线缝在被子里了。小两口加油，祝你们的爱情也永恒！

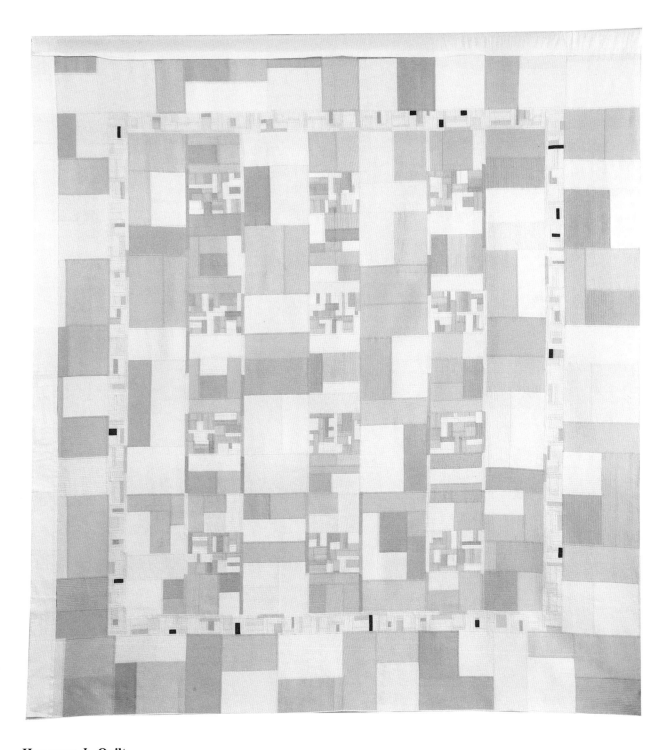

Harmony Ⅰ, Quilt

Silk, 253cm×236cm, Quilting, 1990

  The quilt for the eldest son and his wife that took three years to complete. I have deposit my whole life in this colorful cloth and thread, and tell my trivial things and love with these colorful threads. The colorful threads decorated my life. This quilt with colorful fabric and threads showed mother's love poured into. I hope that you will love each other and hope that you will be healthy and happy. I hope the young couple always respect, trust and help each other in a happy family, and always love each other like a patching cloth. I hope that every member of the family will live in harmony with nature. I put the best wishes in the quilt with a needle. I wish your love will last forever!

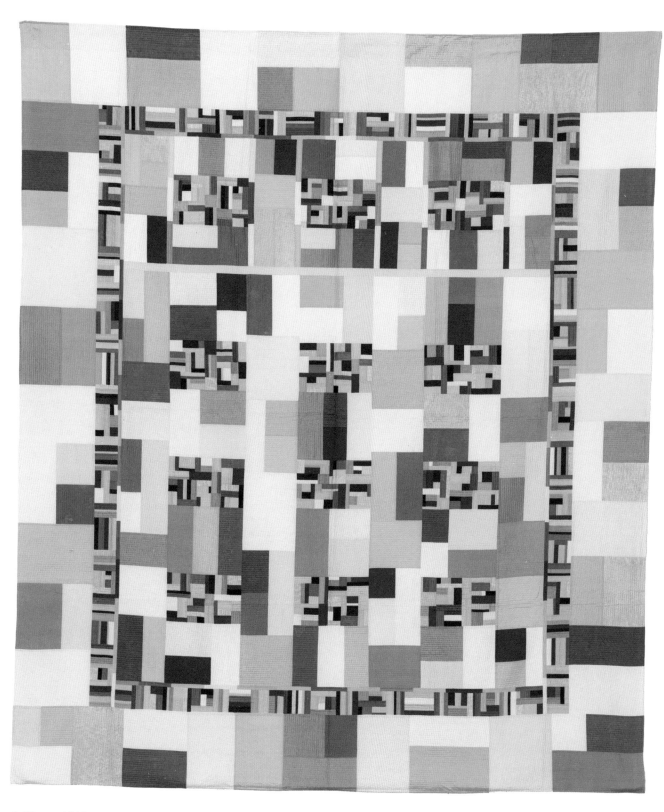

和谐 II 棉被

绸，240cm×210cm，绗缝，1998 年

这是给二儿子、二儿媳制作的被，耗时五年完成。

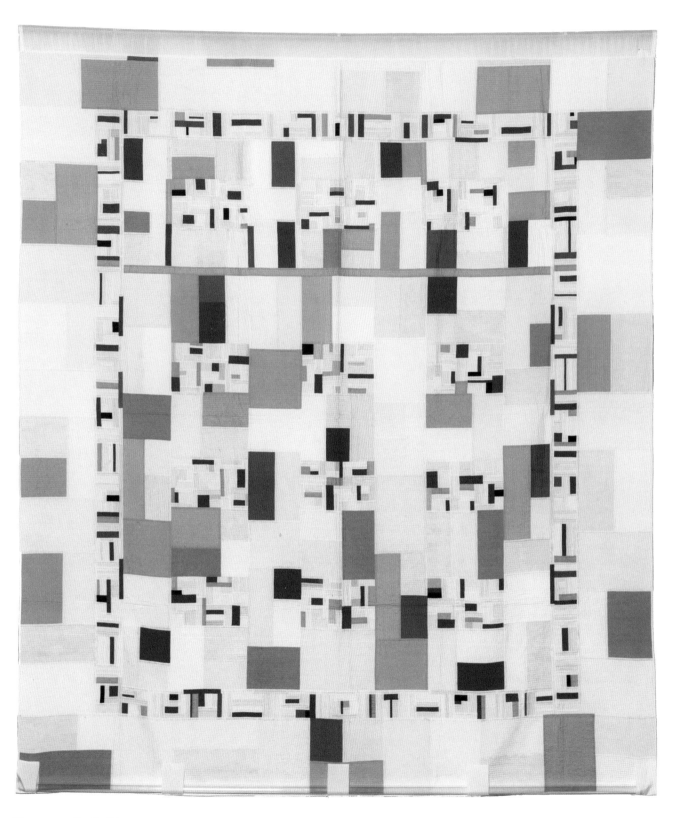

Harmony Ⅱ, Quilt

Silk, 240cm×210cm, Quilting, 1998

It was made for the second son and his wife which took five years to complete.

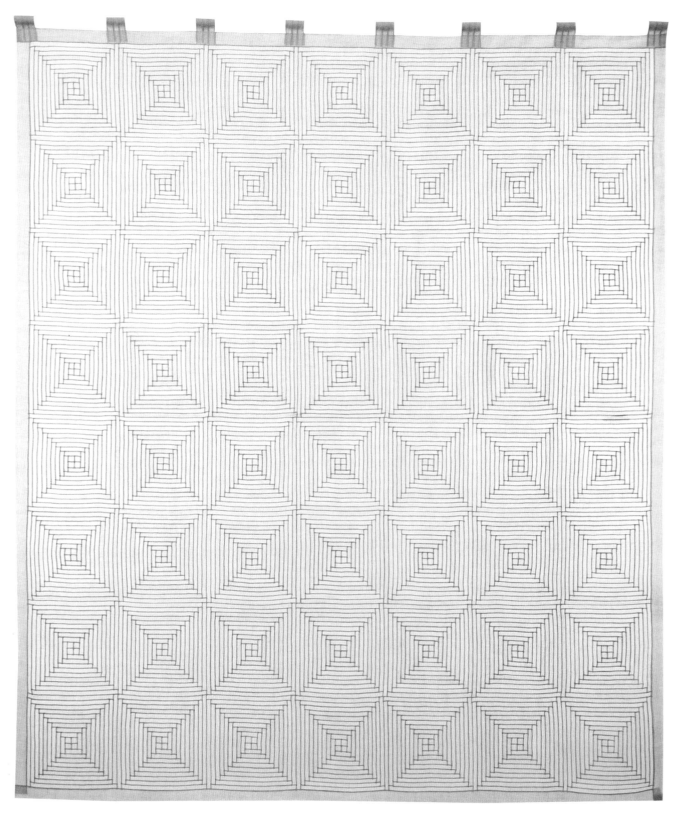

吉祥如意 I　窗帘

苎麻，200cm×180cm，植物染（蓝靛和红花）、撩针、握手缝，1994 年

为大儿子结婚制作的窗帘。

Good Luck I , Curtain

Ramie, 200cm×180cm, Plant dyeing (Indigo and safflower), Slip stitch & Flat felled seam, 1994

This Curtain is made for the wedding of my eldest son.

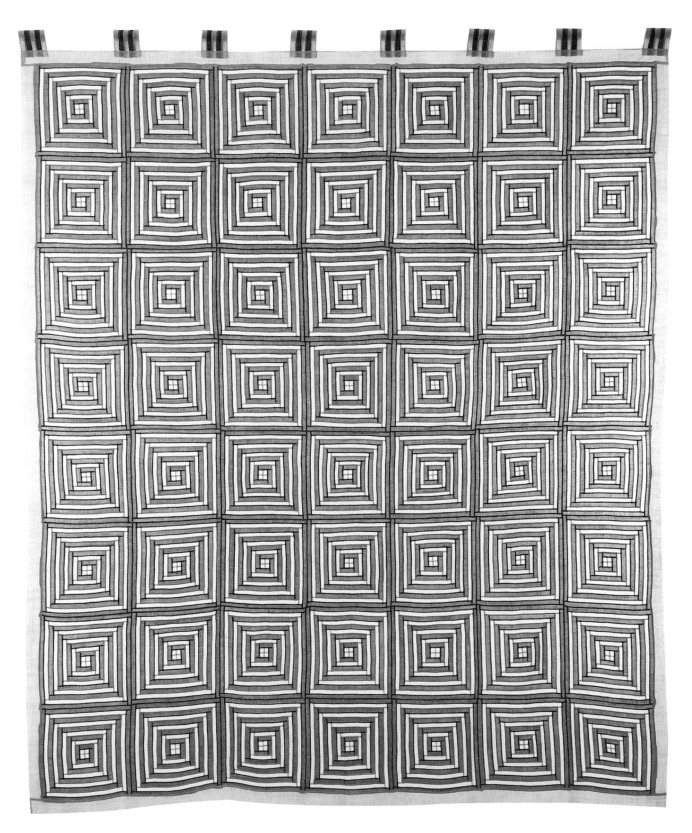

吉祥如意 II　窗帘

苎麻，190cm×160cm，植物染（蓝靛）、撩针、握手缝，1995 年

为二儿子结婚制作的窗帘。

Good Luck II, Curtain

Ramie, 190cm×160cm, Plant dyeing (Indigo), Slip stitch & Flat felled seam, 1995

This curtain is made for my second son's marriage.

(局部 Details)

033

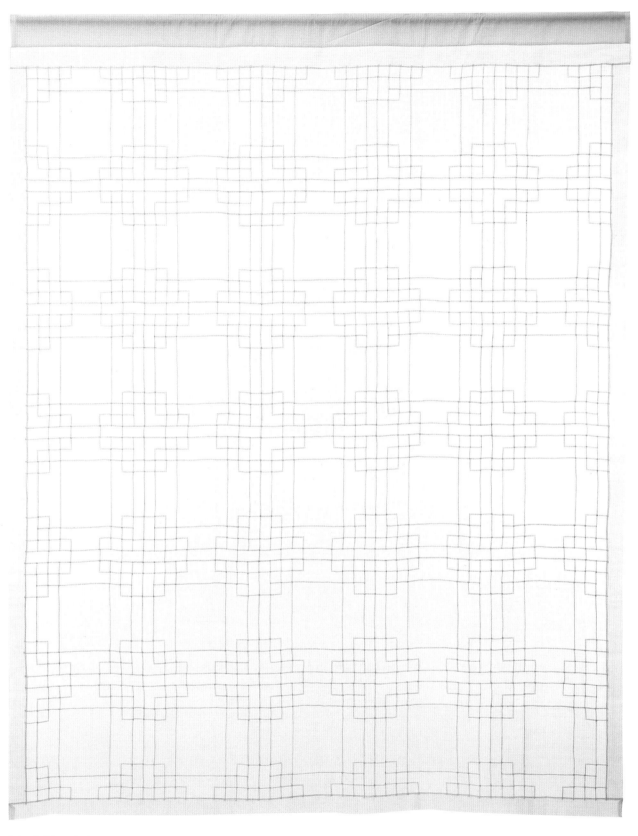

窗棂之美　窗帘

苎麻，210cm×170cm，撩针、握手缝，1995 年
图案来源于江苏苏州山塘街玉涵堂

Beauty of window lattice, Curtain

Ramie, 210cm×170cm, Slip stitch & Flat felled seam, 1995
Patlern from Yuhantang, Shantang street, Suzhou, Jiangsu

(局部 Details)

天外天　窗帘

苎麻，190cm×170cm，植物染（柿子）、撩针、握手缝，1996 年

  该作品以苎麻为原料制作，用柿子染制而成。不同的染色次数，呈现出不同的黄褐色调。柿蒂染出的色彩，还可以呈现黑绿色。透过这扇窗帘，作者想到的是，传统社会朝鲜族妇女大多数都会放弃事业而全身心投入到家庭中。但是，人外有人、天外有天，她们对梦想的渴望与追求并没有因此而停滞。将每一针每一线的缝制都当成向梦想走近的每一步。作品体现了女人简朴生活的美德以及女人独有的气质与追求。

Outer heaven, Curtain

Ramie, 190cm×170cm, Plant dyeing & Slip stitch, 1996

The work is made from ramie and dyed with persimmon. Different staining times, showing different yellow-brown tones. The color of the persimmon calyx, it also appears black and green. This curtain reflects the author's thought that most of the Korean women in the traditional society gave up their careers and devoted themselves to the family. Outside the human has the human, and outside heaven has the heaven. The desire for dreams and pursuits has never stopped. A stitch in the sewing is a step closer to the dream. The work reflects the virtues of women's simple life and their unique temperament and pursuit.

(局部 Details)

(局部 Details)

窗帘

苎麻，170cm×170cm，撩针、握手缝，1994 年

该窗帘是用化学染料手工染的。

Curtain

Ramie, 170cm×170cm, Slip stitch & Flat felled seam, 1994

This contain is hand dyed with chemica dyes.

(局部 Details)

吉祥如意Ⅲ　窗帘

苎麻，300cm×220cm，植物染（红花）、撩针、握手缝，2008 年

这幅作品远看呈现红色的柿蒂纹图案，是用红花染制而成的。但是仔细看每个图案单元的中心，其实是一个万字纹即"卍"。它是吉祥的符号，一圈一圈的字层层扩展开来，象征生生不息。这幅窗帘是2007年为儿子乔迁新居所做，以此祝福儿孙们在生活中诸事顺利、吉祥如意。

Good Luck Ⅲ , Curtain

Ramie, 300cm×220cm, Plant dyeing (Safflowe), Slip stitch & Flat felled seam, 2008

From a distance, this patchwork looks like red persimmon calyx patterns, which is dyed with safflower. Look carefully at the center of each pattern unit, It is actually a swastika pattern, and expanding the circle by circle, is a symbol of auspiciousness. This curtain was made in 2007 for my son to move to a new home, which is used as a wish that the children and grandchildren has a smooth and auspicious life.

(局部 Details)

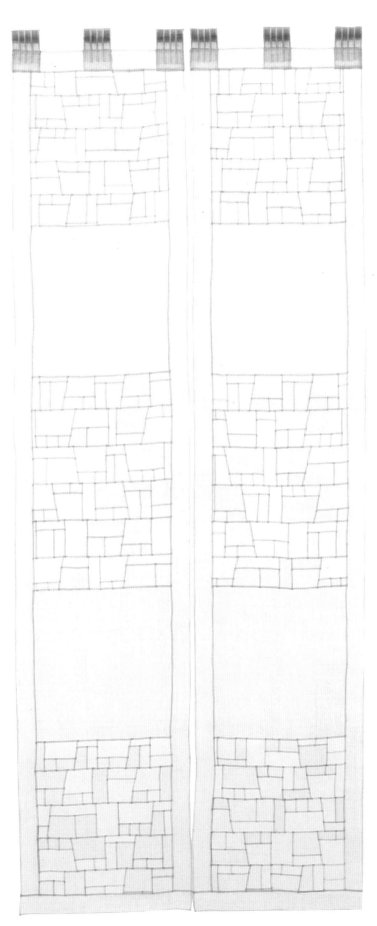

隔断帘

苎麻，172cm×36cm/幅，撩针，1997年

Partition curtain

Ramie, Each, 172cm×36cm, Slip stitch, 1997

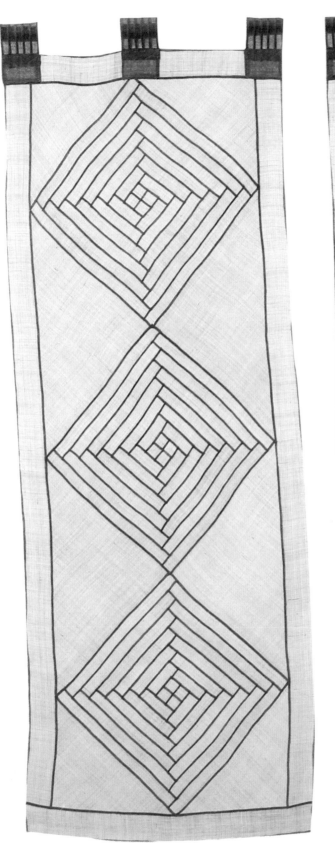

隔断帘

苎麻，130cm×50cm/幅，植物染、撩针，1997年

Partition curtain

Ramie, Each, 130cm×50cm, Plant dyeing & Slip stitch, 1997

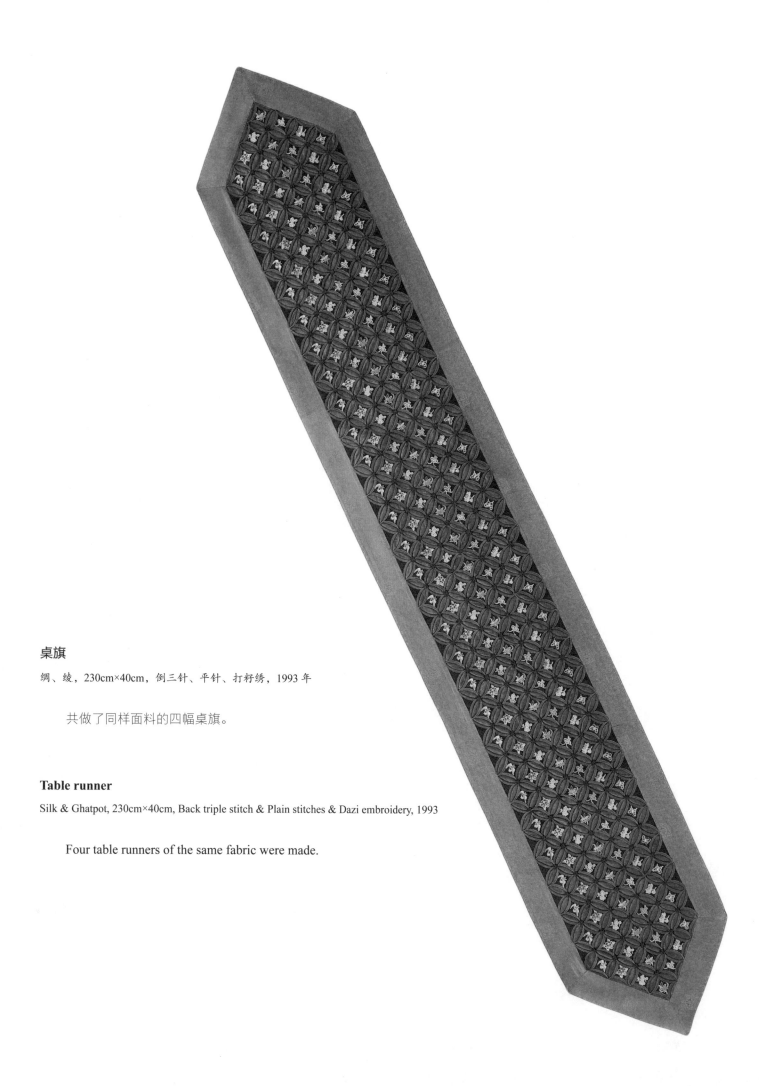

桌旗

绸、绫，230cm×40cm，倒三针、平针、打籽绣，1993 年

共做了同样面料的四幅桌旗。

Table runner

Silk & Ghatpot, 230cm×40cm, Back triple stitch & Plain stitches & Dazi embroidery, 1993

Four table runners of the same fabric were made.

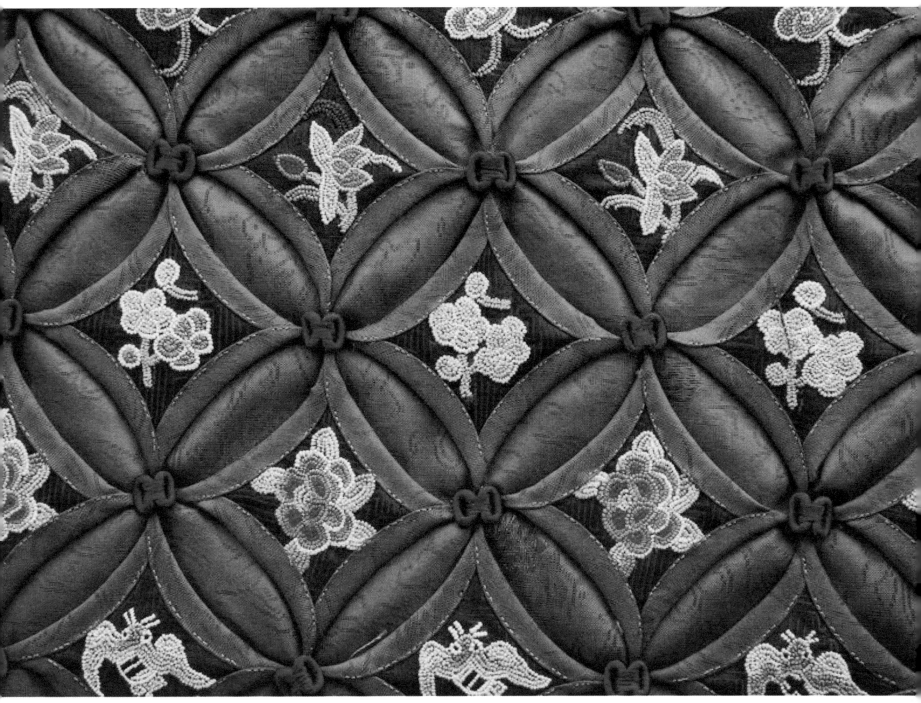

(局部 Details)

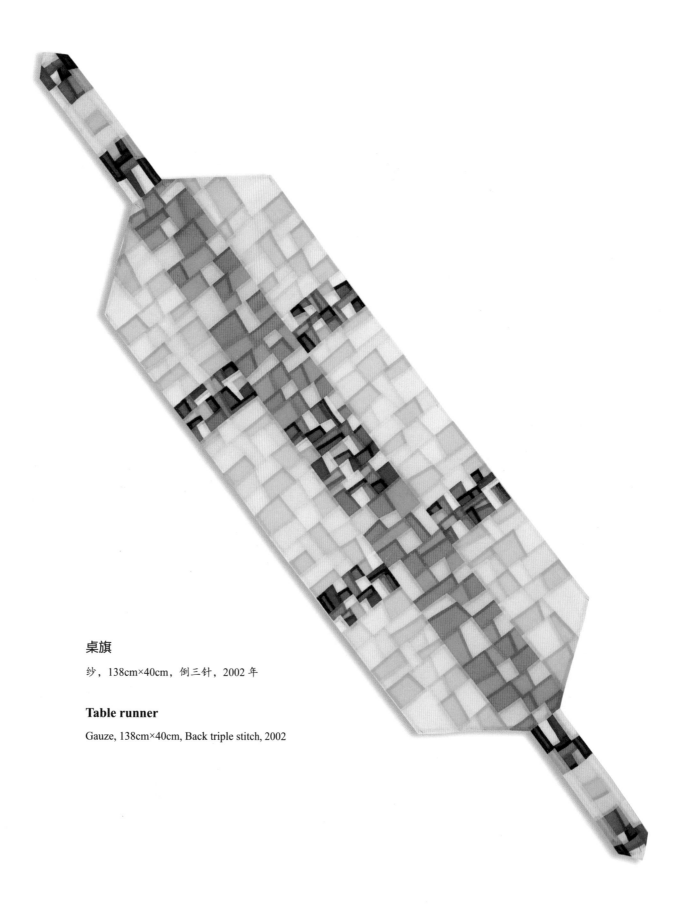

桌旗

纱,138cm×40cm,倒三针,2002 年

Table runner

Gauze, 138cm×40cm, Back triple stitch, 2002

(局部 Details)

053

桌旗与餐垫

纱，桌旗：422cm×180cm；餐垫：40cm×40cm，倒三针、撩针，1998 年

Table runner & Dish cushions

Gauze, Table runner: 422cm×180cm; Dish cushions: 40cm×40cm, Back triple stitch & Slip stitch, 1998

(局部 Details)

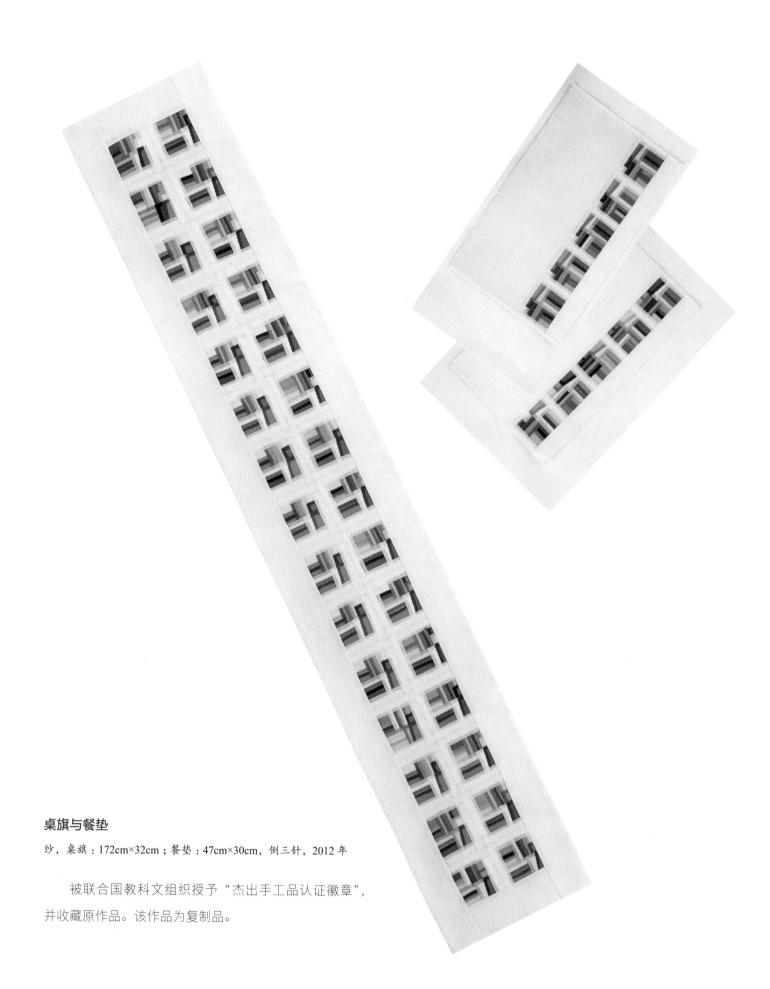

桌旗与餐垫

纱，桌旗：172cm×32cm；餐垫：47cm×30cm，倒三针，2012 年

被联合国教科文组织授予"杰出手工品认证徽章"，并收藏原作品。该作品为复制品。

(局部 Details)

Table runner & Dish cushions

Gauze, Table runner: 172cm×32cm; Dish cushions: 47cm×30cm, Back triple stitch, 2012

Awarded the "Outstanding Handicraft Certification Badge" by UNESCO. The original work was collected by UNESCO, this work is a reproduction.

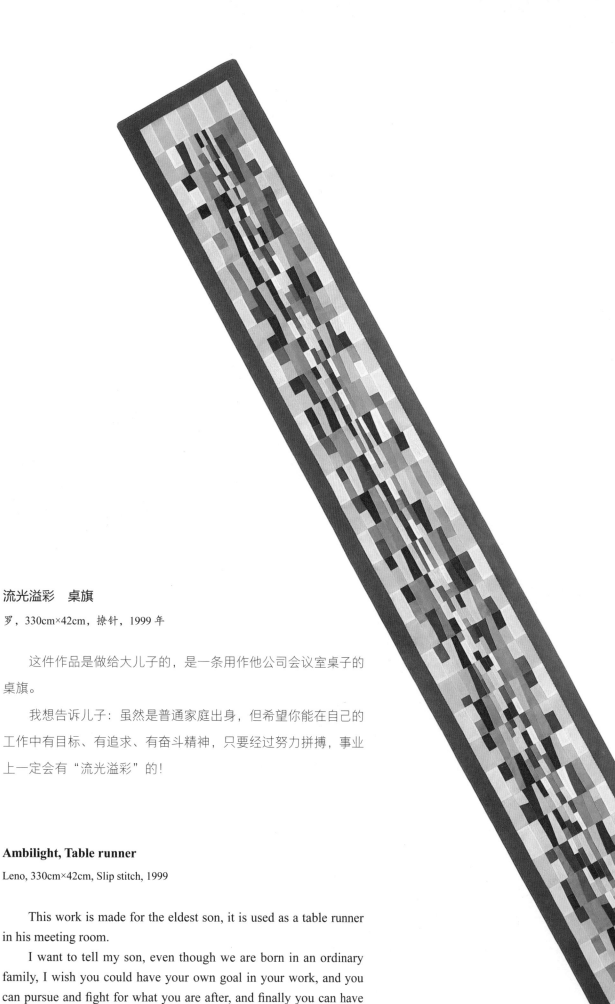

流光溢彩　桌旗

罗，330cm×42cm，撩针，1999 年

这件作品是做给大儿子的，是一条用作他公司会议室桌子的桌旗。

我想告诉儿子：虽然是普通家庭出身，但希望你能在自己的工作中有目标、有追求、有奋斗精神，只要经过努力拼搏，事业上一定会有"流光溢彩"的！

Ambilight, Table runner

Leno, 330cm×42cm, Slip stitch, 1999

This work is made for the eldest son, it is used as a table runner in his meeting room.

I want to tell my son, even though we are born in an ordinary family, I wish you could have your own goal in your work, and you can pursue and fight for what you are after, and finally you can have your "Ambilight" moment in your life.

（局部 Details）

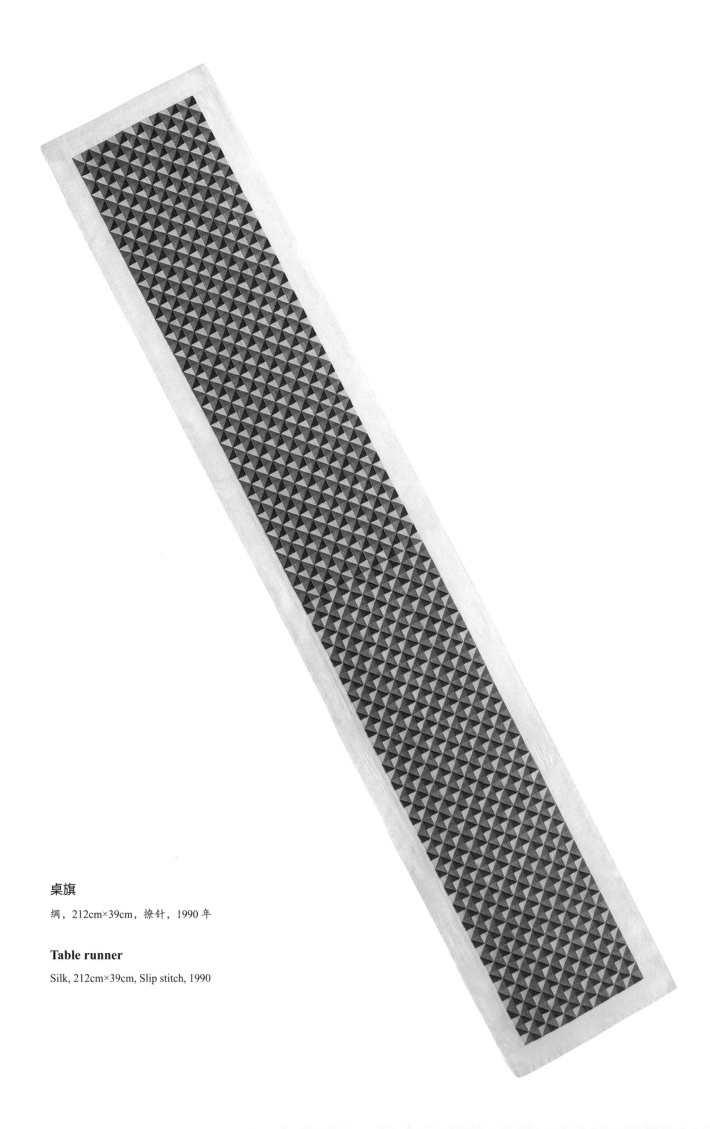

桌旗

绸，212cm×39cm，撩针，1990 年

Table runner

Silk, 212cm×39cm, Slip stitch, 1990

(局部 Details)

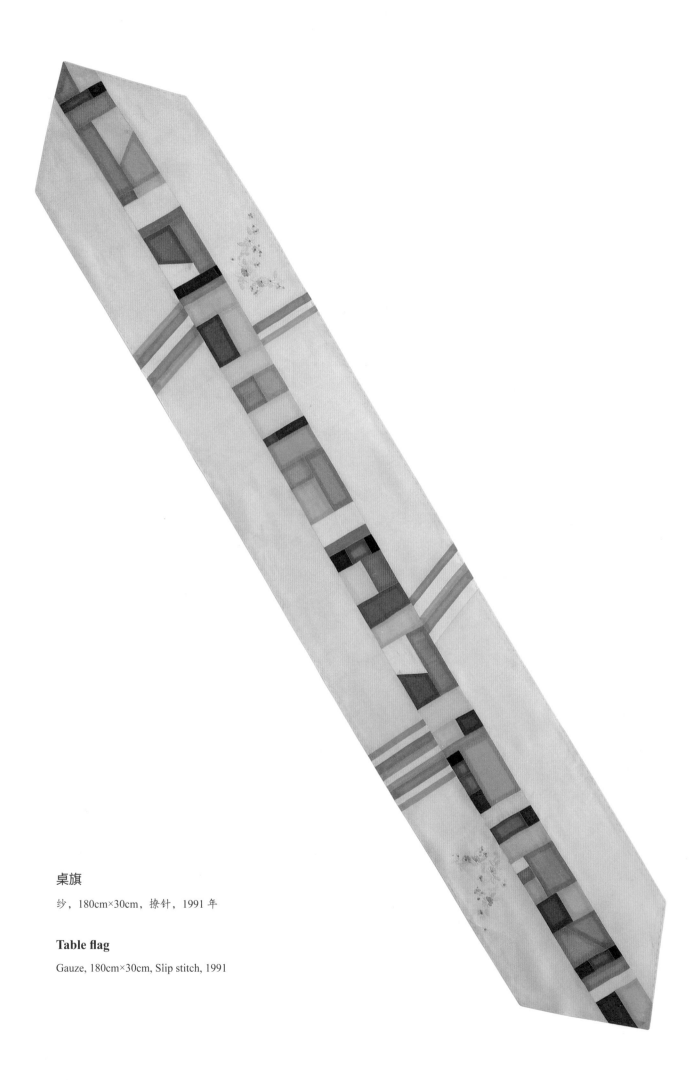

桌旗

纱,180cm×30cm,撩针,1991 年

Table flag

Gauze, 180cm×30cm, Slip stitch, 1991

(局部 Details)

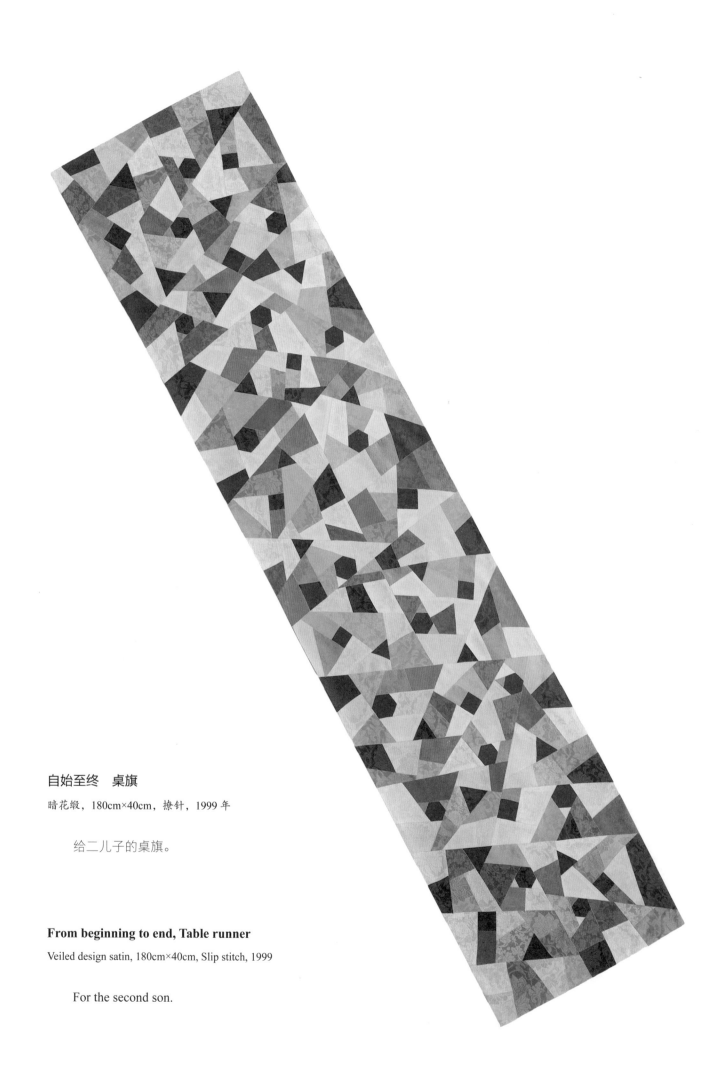

自始至终　桌旗

暗花缎，180cm×40cm，撩针，1999 年

给二儿子的桌旗。

From beginning to end, Table runner

Veiled design satin, 180cm×40cm, Slip stitch, 1999

For the second son.

(局部 Details)

茶盘垫

暗花纱，29cm×29cm，撩针，1998 年

Tea tray pads

Veiled design gauze, 29cm×29cm, Slip stitch, 1998

（局部 Details）

二、服饰
Garment

穿在身上的拼布服饰最能给人留下过目不忘的印象。朝鲜族的传统服装运用了拼布的技法，经过改良的拼布服装也无意中引领了时尚潮流。给小孩子做"百衲衣"更是体现了浓浓的关爱。即使是一条围巾、一个手包、一对耳环，运用了拼布技艺，就会显得颇有特色。身上的穿着，是"女人的香气"的最佳代言。

The patchwork garment wearing on the body will leave an unforgettable impression. The traditional garment of the Korean people use the technique of patchwork, and the improved patchwork garment also leads the fashion trend inadvertently. The "Bainayi" for children reflects deep love. Even a scarf, a handbag or a pair of earrings, using the patchwork technique, will look quite distinctive. The garment on the body is the best endorsement of "woman's aroma".

母亲的香气 — 金媛善拼布布艺术

朝鲜族传统大衣

罗,撩针,1990 年

Traditional female robe of Korean

Leno, Slip stitch, 1990

072

(局部 Details)

朝鲜族改良女衣裙

上衣：绸，裙子：罗，倒三针、平针，1998年

本套拼布衣裙已捐赠给北京服装学院民族服饰博物馆永久收藏。

Reformative female jacket and skirt of Korean

Jacket: Silk, Skirt: Leno, Back triple stitch & Plain stitch, 1998

This suit was donated to the BIFT Costume Museum as a permanent collection item.

(局部 Details）

女式大衣

绸、棉花，植物染、绗缝，2002 年

Female coat

Silk & Cotton padding, Plant dyeing & Quilting, 2002

(局部 Details)

女式背心

绸，撩针，1997 年

Female vest

Silk, Slip stitch, 1997

(局部 Details)

朝鲜族改良女衣裙

上衣：苎麻，裙子：绫，撩针、握手缝，2000 年

Reformative female jacket and skirt of Korean

Jacket: Ramie, Skirt: Ghatpot, Slip stitch & Flat felled seam, 2000

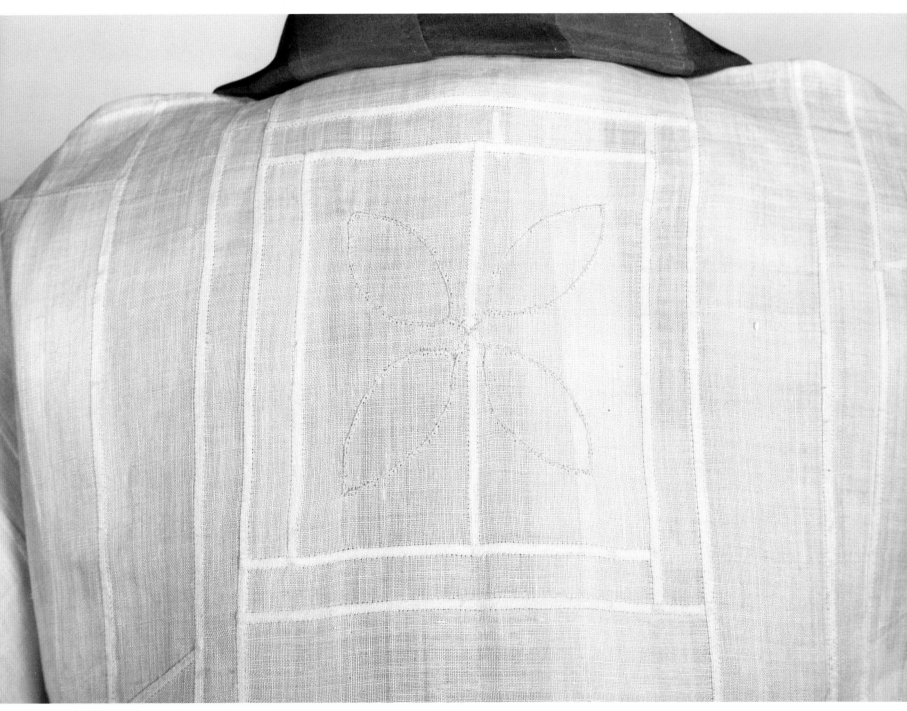

(局部 Details)

背心裙

绸、棉花，绗缝、茶叶染，1989 年

Vest dress

Silk & Cotton padding, Quilting & Tea dyeing, 1989

(局部 Details)

背心 — **Vest**

绸，撩针，2002 年 — Silk, Slip stitch, 2002

(局部 Details)

大衣

纱,平针、撩针、倒三针,2012 年

Coat

Gauze, Plain stitch & Slip stitch & Back triple stitch, 2012

(局部 Details)

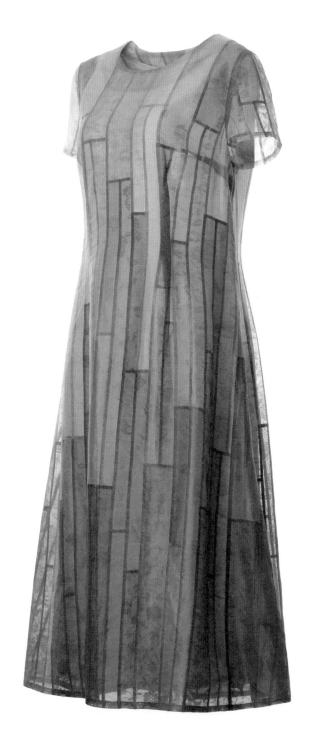

连衣裙

暗花纱,平针,2002 年

Dress

Veiled design gauze, Plain stitch, 2002

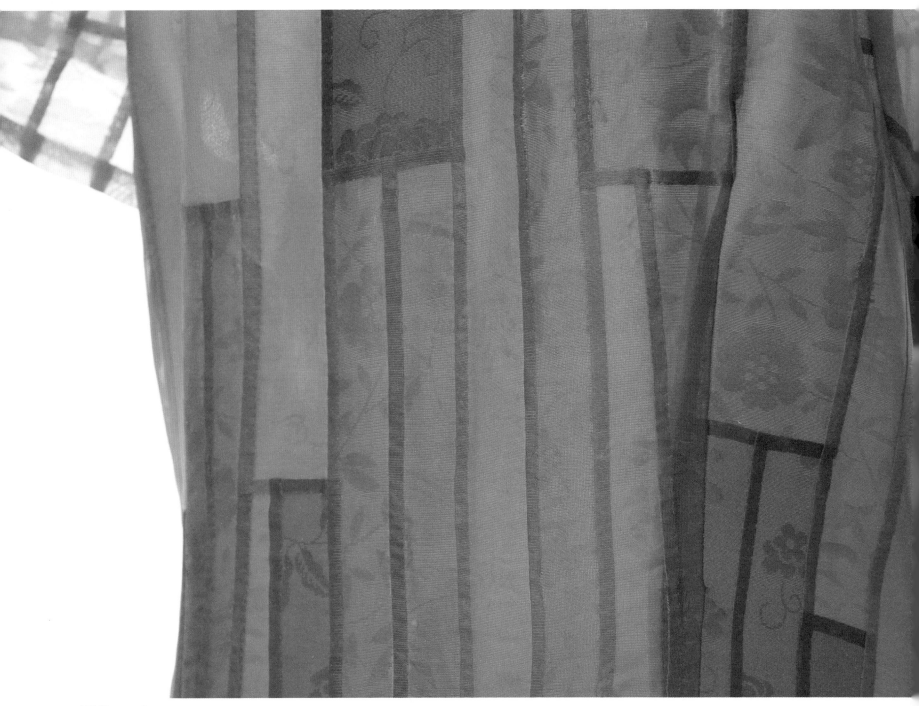

（局部 Details）

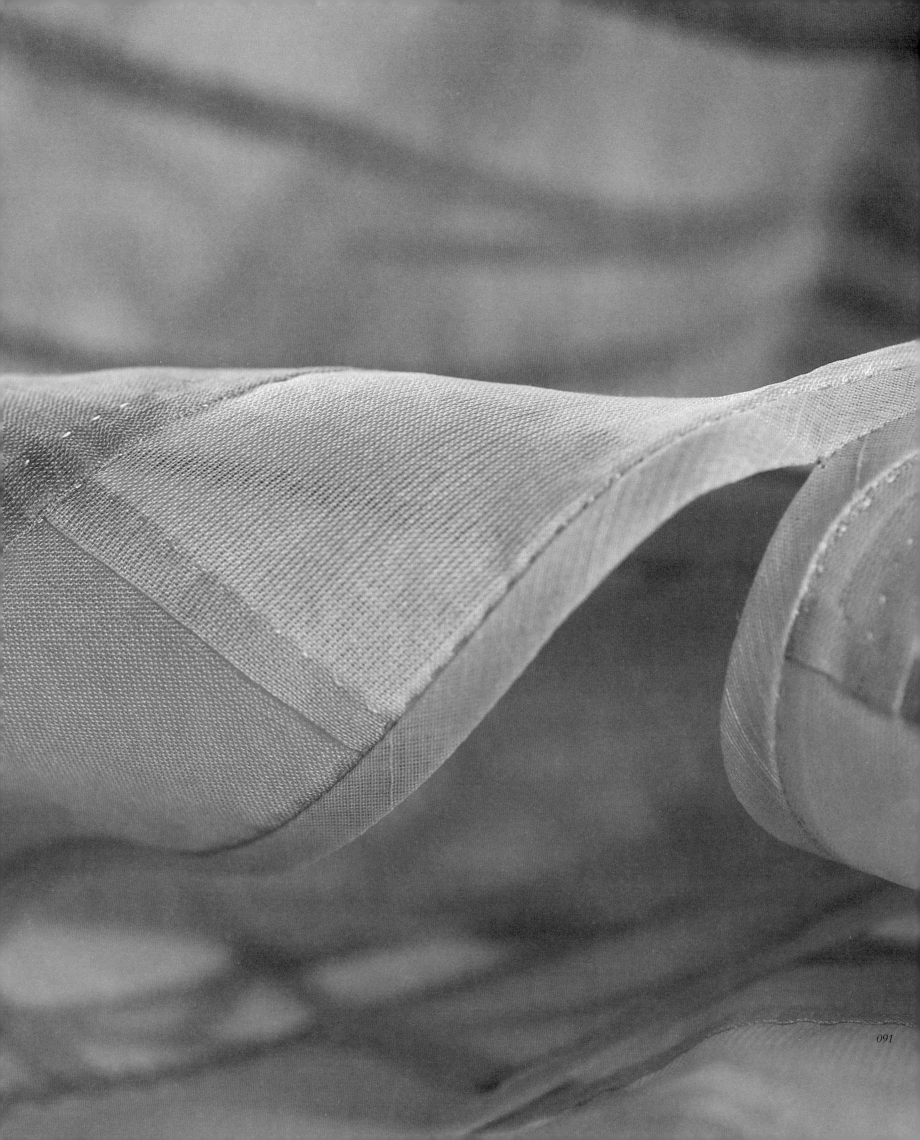

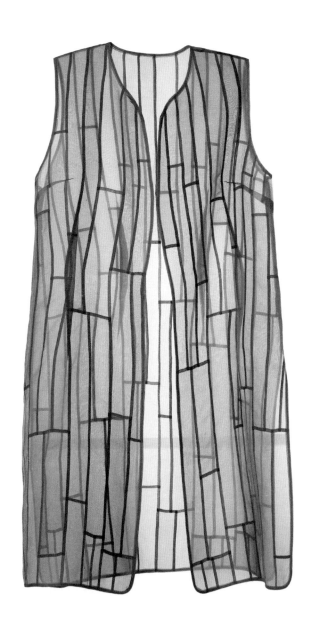

背心与开衫

纱,平针、握手缝,2011 年

Vest & Shirt

Gauze, Plain stitch & Flat felled seam, 2011

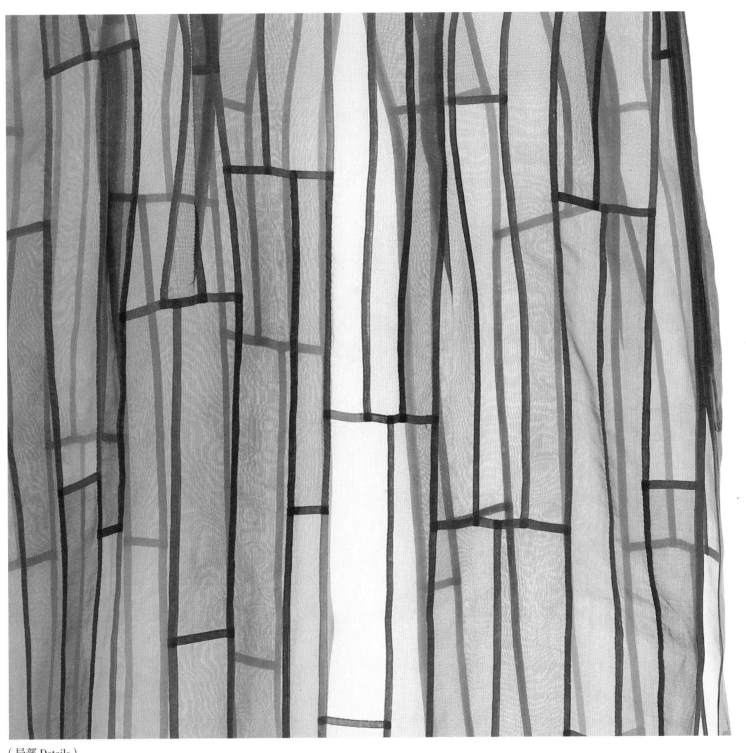

(局部 Details)

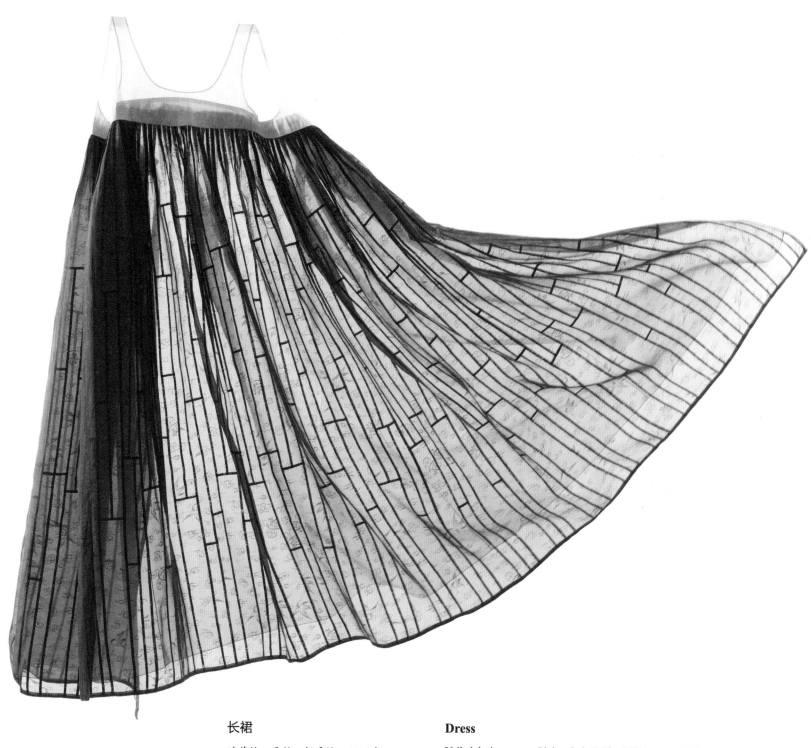

长裙

暗花纱，平针、握手缝，1998 年

Dress

Veiled design gauze, Plain stitch & Flat felled seam, 1998

(局部 Details)

僧袍

宣纸，平针、绗缝，1989 年

将宣纸经过墨染后，绗缝制作而成。这是我 1988 年在韩国看到一位有名的和尚所穿着的衲衣，而后仿照其样式制作的。

Ragged robe

Xuan paper, Plain stitch & Quilting, 1989

The Xuan paper is dyed by Chinese ink and made by quilting. This is what I saw in 1988 in Korea, where a famous monk wore a dress, and then imitated it.

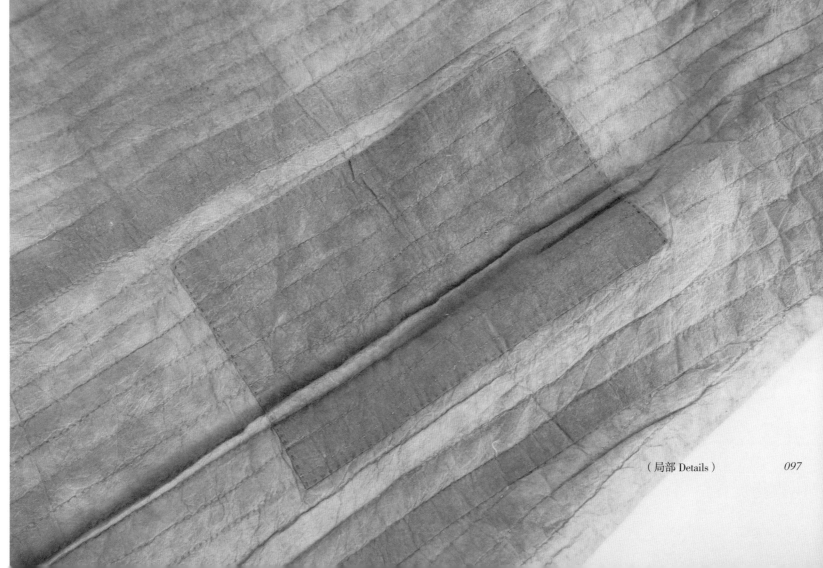

(局部 Details)

孙子的"百天衣"

绸、棉花,绗缝,1999 年

Grandson's suit for one-hundred days old ("Baitianyi")

Silk & Cotton padding, Quilting, 1999

母亲的香气—金媛善拼布艺术

(局部 Details)

孙子、孙女的"出生服"
绸、棉花，绗缝、倒三针，2001 年

New born baby's clothing
Silk & Cotton padding, Quilting & Back triple stitch, 2001

(局部 Details)

朝鲜族女童套装

绸、棉花，绗缝，2000 年

Girl's jacket and skirt

Silk & Cotton padding, Quilting, 2000

(局部 Details)

孙女的衣裙

绸、棉花，绗缝，2000年

Girl's jacket and skirt

Silk & Cotton padding, Quilting, 2000

(局部 Details)

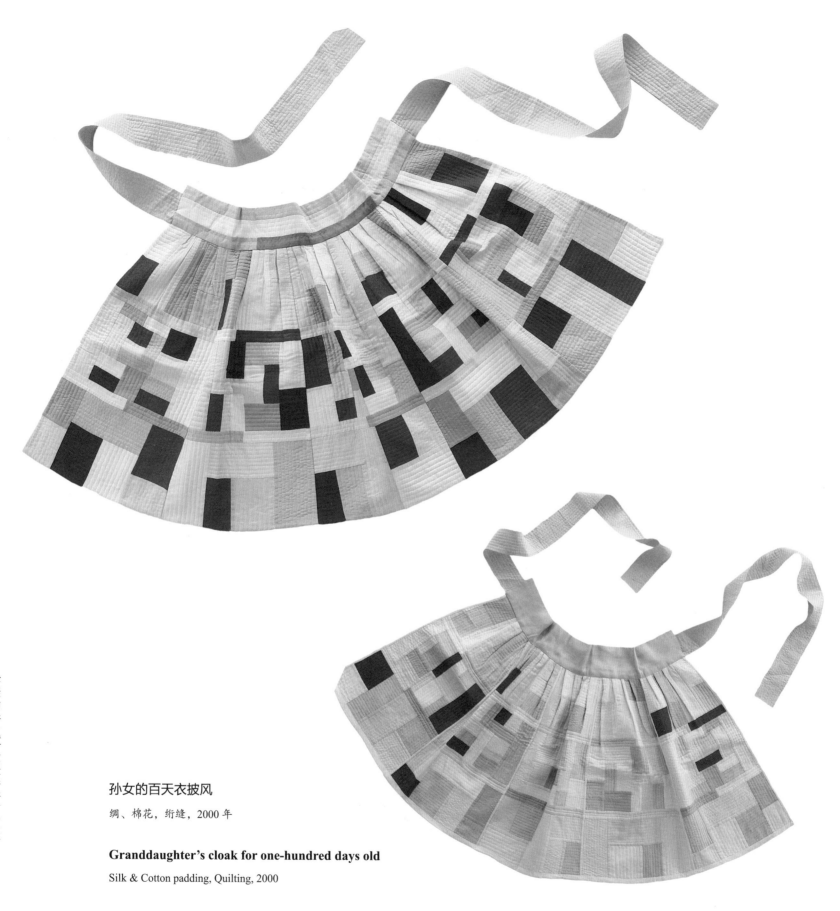

孙女的百天衣披风

绸、棉花，绗缝，2000 年

Granddaughter's cloak for one-hundred days old

Silk & Cotton padding, Quilting, 2000

(局部 Details)

披肩

纱,180cm×20cm,平针、撩针、植物染,
2005 年

Tippet

Gauze, 180cm×20cm, Plain stitch & Slip stitch, Plant dyeing, 2005

围巾

香云纱,180cm×20cm,撩针,2010 年

Scarf

Gambiered canton gauze, 180cm×20cm,
Slip stitch, 2010

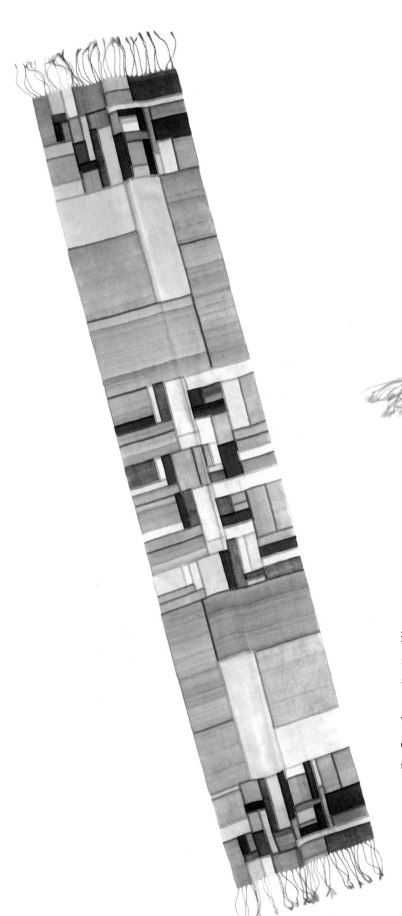

披肩

纱，180cm×20cm（流苏长 16cm），平针、倒三针、握手缝、植物染，2011 年

Tippet

Gauze, 180cm×20cm (macrame: 16cm), Plain stitch, Back triple stitch, Flat felled seam, Plant dyeing, 2011

(局部 Details)

围巾

暗花缎，180cm×20cm，撩针，1996 年

Scarf

Veiled design satin, 180cm×20cm, Slip stitch, 1996

(局部 Details)

围巾

暗花缎，180cm×20cm，撩针，1996 年

Scarf

Veiled design satin, 180cm×20cm, Slip stitch, 1996

(局部 Details)

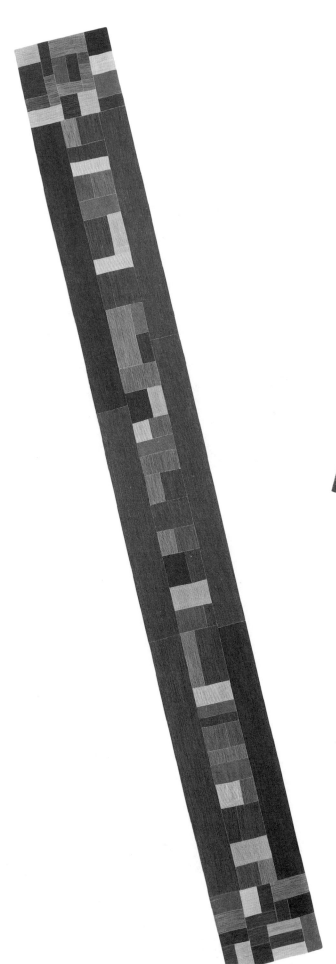

围巾

绨，180cm×20cm，撩针、植物染，1995 年

Scarf

Bengaline, 180cm×20cm, Slip stitch, Plant dyeing, 1995

(局部 Details)

围巾

暗花缎，180cm×20cm，撩针，1995 年

Scarf

Veiled design satin, 180cm×20cm, Slip stitch, 1995

(局部 Details)

围巾

丝绵混纺斜纹绸，129cm×20cm，倒三针，2002 年

Scarf

Cotton-silk twill, 129cm×20cm, Back triple stitch, 2002

(局部 Details)

围巾

绸、暗花纱，180cm×20cm，倒三针，1991 年

Scarf

Silk & Veiled design gauze, 180cm×20cm, Back triple stitch, 1991

(局部 Details)

围巾

暗花缎，180cm×20cm，撩针，1992 年

Scarf

Veiled design satin, 180cm×20cm, Slip stitch, 1992

(局部 Details)

(局部 Details)

围巾

绸，180cm×37cm（流苏长 16cm），平针，1990 年

Scarf

Silk, 180cm×37cm (macrame: 16cm), Plain stitch, 1990

（局部 Details）

母亲的香气—金媛善拼布艺术

挎包

纱,撩针,2010 年

Bag

Gauze, Slip stitch, 2010

挎包

暗花纱，倒三针，2010 年

Bag

Veiled design gauze, Back triple stitch, 2010

手包

纱，倒三针，2001 年

Handbag

Gauze, Back triple stitch, 2001

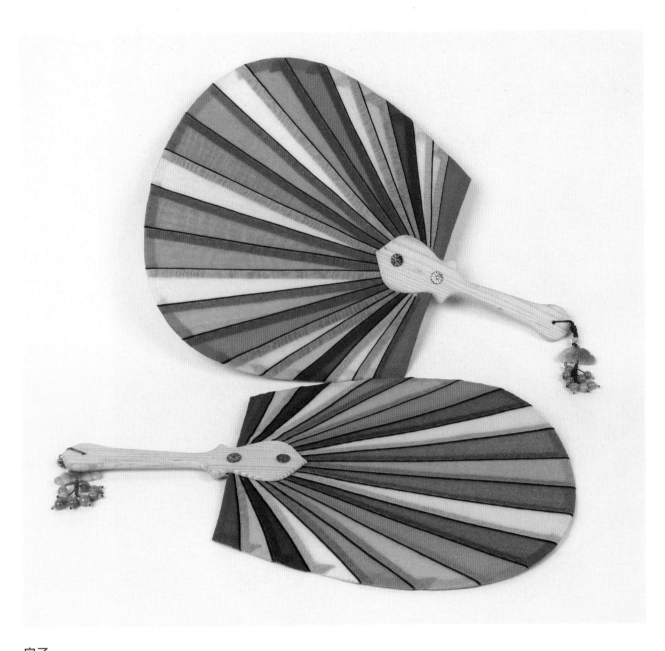

扇子

纱,倒三针,2001 年

Fans

Gauze, Back triple stitch, 2001

母亲的香气—金媛善拼布艺术

138

首饰

纱,卷布,2000 年

Accessories

Gauze, Hemming, 2000

三、壁挂
Wall-hanging

拼布从女性们善意的手工荣升到艺术的殿堂，母亲们也成为了艺术家，实现了更高的社会价值。这里呈现的壁挂作品，从生活中来而高于生活，均为大幅作品。每件作品都耗费我一年甚至几年的心血，凝结了我对于生活的感悟，是我的拼布技艺的集中体现。

The patchwork has stepped into the pantheon of arts from the graceful handcraft of women, the mothers have become artists and achieved higher social values. The wall-hanging works presented here are from life and higher than life. They are all large works. Each piece of work takes me a year or even several years of hard work, which condenses my feelings about life and my patchwork skills.

百花争艳　壁挂

绡，200cm×180cm，叠花、平针、倒三针、撩针，2002 年

第七届亚洲纤维艺术展参展作品、第九届"色彩中国"2017年度中国传统工艺色彩奖（拼布类）

Hundreds Flowers Blossoming, Wall-hanging

Raw silk, 200cm×180cm, Folded flowers, Plain stitch, Back triple stitch & Slip stitch, 2002

The 7th 2017 Asia Fiber Art Exhibition, The 9th 2017 Chinese Traditional Craft Color Award (Patchwork Category)

"太阳花"，是我看到贵州施洞苗族的堆绣之后进行的创作尝试。

1992年，我第一次到贵州施洞时发现了当地苗族衣领背后的堆绣。仔细观察她们制作堆绣的技法，我当时的心情真的无法用语言表达……只是心里想：我一定要用像她们那样的匠人精神来做拼布！

回到家后，我就开始不断地尝试、不断地实践，决定选择用同心圆重叠的方式来制作"太阳花"——这个从研究到实现过程用了将近一年的时间。最后实现，我选用了我常用的真丝绡，不经过皂角水上浆，直接折叠，制成了"太阳花"，这是一个从模仿中获得启发、最后改变创作思路的过程。

整幅《百花争艳》作品，灵感来源于我们的春节家宴。临近钟声敲响的时候，烟花齐放、家人齐聚一堂，我希望这团圆、盛大的瑞象能一直延续下去。于是，我便决定将太阳花排列组合成一幅烟花盛放的图景，从一个单元，到一个长条组合，再到一个方形的组合……最终完成。

这个实现的过程，用了整整七年的时间。这期间，布料的搜集耗费了相当大的精力。每次去韩国的时候，我会从韩服服装店扔掉的废材料中挑选我想要的色彩的面料。但每次捡回来的废料布头始终满足不了我最想要的色彩搭配。前前后后精心地选料、配色、制作、组装……整整七年才完成这幅令自己满意的《百花争艳》。在制作过程中我悟到了一个道理：不是我选择了"拼布"，而是"拼布"选择了我。

拼布的世界是一个有温度的世界，拼布世界给了我生命之养，拼布是我的情感寄托，每当我感到沮丧或身体不适时，只有在五彩的拼布布头中挑选我想要的色彩时会让我重振精神。拼布是我人生中最大的精神支柱！

我对自己的手艺有一种近似于自负的自尊心，对它要求非常苛刻，并为此不厌其烦、不惜代价，但求做到精益求精、完美再完美。

我要求自己一定要从内心热爱拼布、热爱生活；要耐得住寂寞，把孤独当作享受——实际上我真的非常享受这种做拼布时的"孤独的时光"。我想孤独是人生的常态，孤独中更有自我，任思绪放飞。

享受这种外物与自我的相处，就是我秉持的拼布精神——坚守、情怀、修行。

艺术，是一种执着笃定的精神，坚持下去的力量就是情怀。2012年8月，我到中国台湾参加艺术拼布展，展会解说员对我说："加拿大的拼布艺术家看过您的作品《百花争艳》，说这个作品的作者肯定是疯子，不然做不出这么好的作品……"后来，这位加拿大拼布艺术家成了我特别好的朋友，她就是我们"世界灵感八姐妹"之一的玛丽（Mary）。我们经常一起交流，彼此相见恨晚。

我们对于如何让拼布手艺达到熟练精巧有着共同的认知，那就是应该有超乎寻常甚至近乎神经质般的艺术追求。我们对自己的每一件作品，不管再小都力求尽善尽美。对每一件哪怕只是稍加经手的作品也要负起全部责任，否则就是一个拼布人的耻辱。

对于拼布人而言，我认为人品比技术更重要。而做拼布的过程其实也是一种修行的过程：慢慢将内心深处的爱挖掘出来，放在手艺上，如此才能达到一种境界。

《百花争艳》的做法就是重拾失落的拼布人的精神，也是重拾内心的自我。一个手艺人的本质，不外乎两个词：敬业和认真。

(局部 Details)

"Sunflower" was a creative attempt worked by me after I saw the barbola of the Miao nationality of Shi Dong, Guizhou.

In 1992, I found the barbola on the back of the collars from the local Miao nationality when I first went to Shi Dong, Guizhou. After observing the techniques of how they make barbola carefully, I really could not express my feelings in words... I just thought that I will certainly make patchwork in the spirit of craftsmanship like theirs!

After I got home, I started trying and practicing, and I finally chose the method of concentric circles overlap to make "Sunflower" - it took nearly one year from research to implementation. For the final implementation, I chose silk gauze which I used usually, with no starch by soap water, folded it directly to make "Sunflower". This was a process which I was first inspired by an imitation, and I finally changed my creation thoughts.

The inspiration of the work *Hundreds Flowers Blooming* came from our family feast in the Spring Festival. When the stroke of the bell was drawing near, fireworks exploded, families gathered together. I wished this united and grand auspicious phenomenon could go on forever. Therefore, I decided to arrange and combine the sunflowers into a picture of fireworks explosion, from one unit, to one strip combination, and then to a quadrate combination... and then it was finally done.

It took me 7 years to complete the work. During this period, it was also exhausting to collect cloth. Every time I went to Korea, I would pick the fabrics in the colors that I wanted in the abandoned waste materials by the hanbok shops. However, I was still not satisfied with the color matching of the waste cloth I collected every time. After I carefully selected the materials, matched the colors, got it finished, had it combined, 7 year has been gone for me to complete this satisfying work *Hundreds Flowers Blooming*. In the process, I realized one fact-I did not select patchwork, but I was selected by patchwork.

The world of patchwork is a warm world. It gives nourishment to my life. Patchwork is my emotional sustenance. Every time I feel depressed or sick, I could only be refreshed by picking the fabrics in the colors that I like.

Patchwork is the most powerful spiritual pillar in my life!

I am confident or nearly overconfident in my craftsmanship. I am extremely strict with it. I would take great pains at any cost to achieve excellence.

I asked myself to love patchwork and love life from my heart; to endure loneliness and enjoy loneliness - actually I really enjoyed the time when I was making patchworks alone. I believed that loneliness is the normal status of life. We could become more self-conscious in the loneliness and free our thoughts.

To live in harmony with these foreign objects, was the patchwork spirit that I adhered to - persistence, feelings and discipline.

Art is a persistent and determined spirit. The power to keep going is what we called feelings. In August, 2012, I went to China Taiwan to take part in a patchwork art show. The announcer told me: "The Canadian patchwork artist saw your work *Hundreds Flowers Blooming*, and claimed that the author of this work must be crazy otherwise she could not create such a good work..." Later, this Canadian patchwork artist became my best friend. She is also one of our eight sisters of world inspiration, Mary. We always talked a lot with each other. We both felt it was too late to meet each other.

We hold the common recognition about how to improve our techniques, which means we have the same artistic pursuits, which are extraordinary and almost neurotic. We strive for the perfection in every piece of our works, no matter how small they are. We shoulder full responsibility for every piece of our works. Otherwise, it would be a shame for patchwork artists.

For patchwork artists, I personally think that moral quality is more important than techniques. The process of making patchworks is also the process of cultivating oneself. You need to gradually discover the love deep inside one's heart, and put it into your techniques. Only by doing so can you achieve a higher level.

The practice of *Hundreds Flower Blooming* was not only to regain the spirits of the patchwork artists, but also to regain the inner selves. The essence of a craftsman can be described in two words: dedicated and earnest.

(局部 Details)

(局部 Details)

百花齐放　壁挂

绡，200cm×180cm，
卷布、平针，2012 年

"蝙蝠花"，是制作完第一幅《百花争艳》后，对"太阳花"的花型进行的再创作——由中心的极简蝙蝠形向四周不断辐射，生出整个花朵，其中寄寓着满满的"福"意。它的创作灵感，源于我的家人。

第一幅《百花争艳》制作完工后，我兴奋、激动之际想起了我的家人。特别是始终如一的、在背后默默支持着我的姊妹们。

我有四位妹妹和一位可爱的弟弟。家有"五朵金花"——那我就用五种不同的花型来制作一个"百花系列"！

我要用这个系列来表达我对她们的爱，希望她们的生活就像这"蝙蝠花"的含义一样，一生一世，福满爱满。

如今，我已经完成了四种花型的作品，福气不停，创作继续……

(局部 Details)

Hundreds Flowers Flourishing, Wall-hanging

Raw silk，200cm×180cm，Hemming & Plain stitch，2012

"Bat Flower" is the remake of "Sunflower" after my completion of *Hundreds Flowers Blooming*; radiate from the center, the shape of bat, grows the full flower, which means the "fortune" in Chinese language. The idea of this creation comes from my family.

After the completion of *Hundreds Flowers Blooming*, I was extremely excited and happy; and I thought of my family. Especially my sisters always stand behind me and back me up.

I have four sisters and one lovely younger brother. "Five daughters" in the family as five golden flowers, and then I would use five different kinds of flowers to make a flower series.

I would use this series to show my love to them, and I wish their lives were like the "bat flower", fortunate and lucky for all their lives.

Now, I have finished the work with four kinds of flowers, and I will continue to create accompanied with good fortune...

（局部 Details）

(局部 Details)

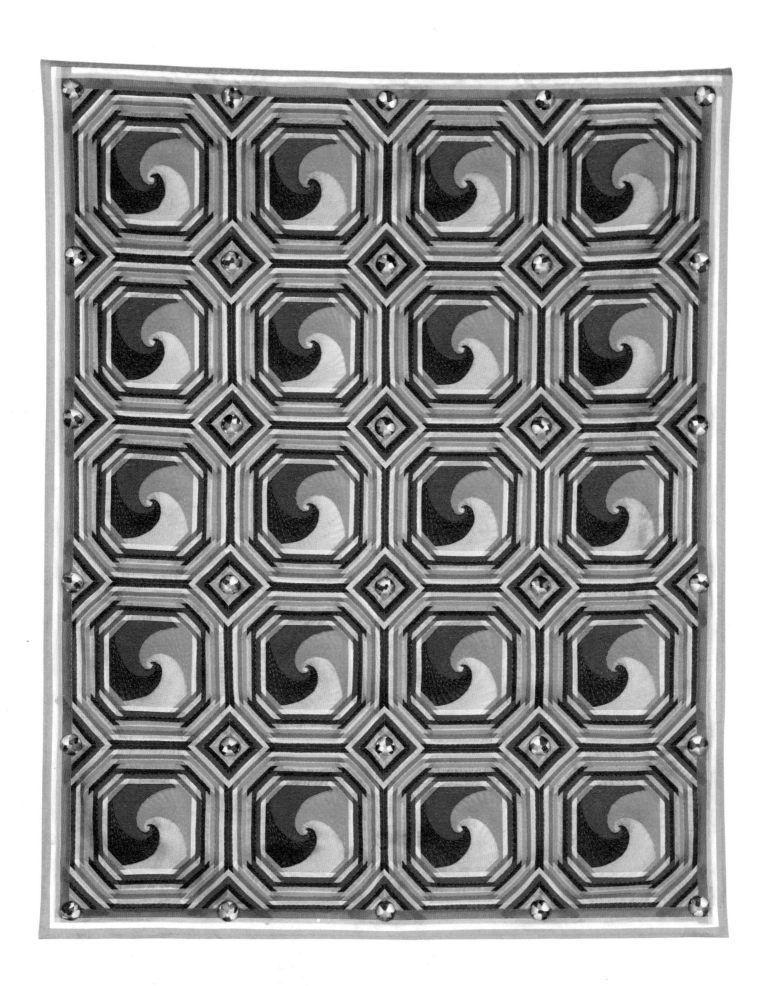

母亲的香气—金媛善拼布艺术

154

念想　壁挂

绡，175cm×145cm，倒三针、撩针，1997年
2008年中国国际家用纺织产品设计大赛银奖

这幅《念想》壁挂，是为了参加韩国展览而做的，所以图案设计灵感来源于韩国的国旗。我是朝鲜族，与韩国语言共通，但是这幅作品在缝制的时候，我用的针法"语言"全部是"倒三针"。

"倒三针"并非拼布的惯用针法，我用这种针法是希望把中国的传统文化融入拼布这种老手艺中去。古人讲："退一步海阔天空"，所以我退一针、退一针、再退一针，然后大跨步前进一步，这就是中国人"以退为进""三思而后行"的理念，这其中包含着我们中华民族的智慧与性格。

我用五彩针插作为中间的装饰，是想要体现女人的香气——每每当我拿起祖母和母亲的老物件时，总能感受到一股香气扑面而来，我认为那就是母亲的香气。

针插同针、线一样，都是女红中不可或缺的工具，女人们日日缝补，它就从旁陪伴、记录下一针一线的日日夜夜与酸甜苦辣，也见证了女人们付诸手艺的情感和爱。此外，中国人善用谐音暗喻吉意，所以我在针插上面又做了蝙蝠结，寓意"福（蝠）在眼前"。

我相信，女人用善心、巧手、大爱和智慧所创造出来的作品是能为观者带来幸福感的。

(局部 Details)

Missing, Wall-hanging

Raw silk, 175cm×145cm, Back triple stitch & Slip stitch, 1997

2008 China International Home Textiles Design Competition Silver Award

  This *Missing* wall-hanging is made to participate in Korean exhibitions, so the design is inspired by the Korean flag. I am a Korean nationality person and share the same language with the Korean. However, the sewn "language" of this piece used was a traditional technique of "three needles".

  Traditional "three needles" is not often seen in patchworks. I applied this way of stitching in patchwork to put Chinese traditional culture into this old craftsmanship. Ancient people said, "Take a step back and you can see the light", so I take a stitch back, another stitch back, and then another, and then step forward, this is the Chinese concept of "retreating as forwarding" "thinking and then going", which includes the wisdom and character of the Chinese nation.

  I use the multicolored pin cushion as the middle decoration; I want to embody the fragrance of a woman. Whenever I pick up the old things of my grandmother and mother, I can always feel a scent. I think that is the mother's smell.

  The pin cushion is the same as the needle and the thread; it is an indispensable tool for the female handwork. Women sew every day. It is accompanied with happiness and sorrow, with women' emotions and love of craftsmanship. Apart from that, Chinese use the homonym to get a good meaning. So I made a Bat Knot on it, which means "good fortune is knocking" in Chinese.

  I believe that works made by women' kind heart, delicate hands, love and wisdom are able to bring blessedness for people.

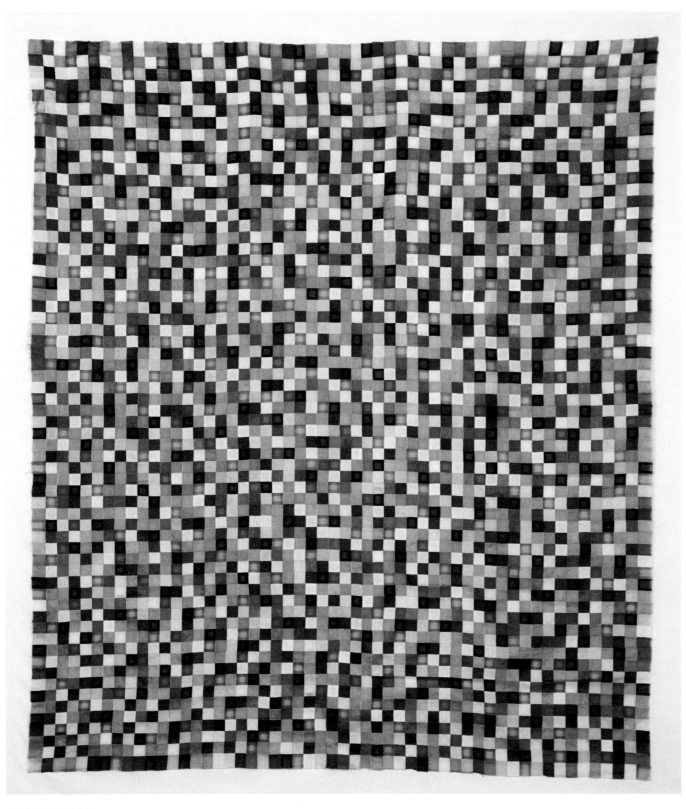

光芒四射　壁挂

绸，180cm×160cm，撩针，2006 年

　　这个作品是用边长 3cm 的小方块布料拼成，颇有马赛克拼贴画的趣味。

(局部 Details)

Radiant, Wall-hanging

Silk, 180cm×160cm, Slip stitch, 2006

    This work is made of small squares of cloth with a length of 3cm on each side, which is quite an interesting Mosaic.

母亲的香气—金媛善拼布艺术

节俭之美　壁挂

罗，186cm×176cm，撩针，2009年

　　《节俭之美》这幅作品是为了参加2010年北京服装学院民族服饰博物馆的个展而特意准备的。当时我想，北京服装学院的学生们毕业后都要被分配到纺织服装行业中去，那么什么样的作品能给他们带来惊喜和思考呢？我听说印染厂打样的样品和服装厂裁剪下来的布头都是被扔到垃圾桶里浪费掉的，于是决定制作由布头组成的《节俭之美》。为了使作品更能体现发心，我便用自己制作其他大幅作品后裁剪下来的、大小不一的三角形布头来制作这一件。所用到的最大三角形布头长约11厘米，最小的仅长约1.5厘米。

　　我希望用这幅作品提醒北服毕业的学生到达各自的工作岗位后，一定要将被丢弃的布头们重新收集起来寄回母校，让在校的学生们可以用这些布头来练习、设计各种作品。在这个过程中，学生们也会感受到中华民族惜物节俭的传统美德，那么所谓"节俭之美"，也就得到了升华。

　　学生们初入社会，在全新的工作生活中会遇到很多"棱角现象"——与周遭的关系似乎像三角形一样，尖锐、锋利、容易受到挫折、伤害。怎样才能把这些复杂的关系处理好呢？就像制作这幅作品的过程一样。我之所以用三角形布头，是因为三角形的拼接更加复杂，需要更大的耐心和坚持、更需要善心和热情，当学会用这样的态度为人处世，将会获得意想不到的美满结果。

　　另外，也希望他们潜心钻研、提升自我，在各自的新岗位上成为技术、品德最优秀的"北服毕业设计师"。

　　那年的11月，时任北京服装学院艺术设计学院的詹炳宏院长（现任北京服装学院副校长）邀请我为该学院大四的学生讲一个月的课——我以前在考大学的时候就很向往进入服装设计学院，时隔半个多世纪，我的愿望实现了！

(局部 Details)

Beauty of Frugality, Wall-hanging

Leno, 186cm×176cm, Slip stitch, 2009

  This work *Beauty of Thrifty* is specially prepared for my individual exhibition at Ethnic Costume Museum of Beijing Institute of Fashion Technology in 2010. At that time, the students of Beijing Institute of Fashion Technology would be assigned to the textile and garment industry after graduation. So what kind of works can bring surprises and reflections to them? I heard that the samples of the dyeing factory and the leftover cloths cut by the garment factory were thrown into the trash can, so I decided to make The Beauty of Thrift. In order to make the work more embossing, I made this piece by using the triangular cloths of different sizes that were cut out after making other large works. The largest piece of triangular cloths used is about 11cm long and the smallest is only about 1.5cm long.

  I hope that this work will remind the students who graduated from Beijing Institute of Fashion Technology to start their jobs; They should re-collect the discarded cloths and send them back to their university so that the students at the school can practice and design various kinds of cloths works. In this process, the students will also feel the traditional virtues of the Chinese nation to save things, so the so-called "frugal beauty" will be sublimated.

  When students first enter the society, they will encounter many "edges" in their new work and life — the relationship with the surrounding seems to be like a triangle, sharp to new flesh, they are vulnerable and easy to get injuries. How do we deal with the complex relationships? Just like making this piece of work. The reason why I use a triangular cloth is that the splicing of triangles is more complicated, requires more patience and persistence, and requires more kindness and enthusiasm. When learning to adopt the attitude, people will get unexpected and satisfactory results.

  Apart from that, I hope them to study hard and keep improving themselves, so that they can become the professional and respected "designers from Beijing Institute of Fashion Technology".

  In November of that year, Zhan Binghong, the Dean of the School of Art and Design of Beijing Institute of Fashion Technology Art then (now the Vice President of the Beijing Institute of Fashion Technology), invited me to give a lecture to the students of the School of Dyeing and Weaving Department for a month — I used to be longing for entering this department long ago, and after half a century, and my dream finally come true.

母亲的香气—金媛善拼布艺术

路 I 壁挂

真丝纱，180cm×160cm，倒三针，1999 年

　　这一幅《路》，既是我无法用语言描述出来的人生之路，也是我所奉行的环保低碳的创作之路。

　　由于母亲是裁缝的缘故，形形色色的五彩布头就是我儿时的玩具。直到现在，每次看到它们的时候也依然会油然生出一种温暖的感觉。

　　这幅作品的用材，是做传统的朝鲜族裙子时剪下的布头，它们很长，就像一条条小路。人一生要经历太多太多，其中的喜怒哀乐实在无法用语言表达。有的时候，也会像布头被扔到垃圾桶里那样受到挫折，幸而我的性格从小就很顽强、不服输。

　　这件作品完成后，整个人都沉浸在幸福的喜悦中——真是功夫不负有心人！我的人生路，就以这样的形式在面前铺陈展开。它像一部电影一样唤醒了我的点滴记忆……我相信，那由一块块布头拼出的、无可比拟的斑斓色彩会给每位观者同样的美妙感觉。

　　《路》的另一个意义，是我一直在践行着的、环保低碳的手艺之路。我们有限的生命之所以会创造出无限的价值，是因为将一点一滴的爱、热情和时间都利用了起来，不浪费自己任何一点微小的感触和灵性，就像用布头一点一点拼出一幅幅能给人带来幸福感的作品，每一块小小的布头都可能在整个作品中起到重大的作用——它们也是应该被珍视、不可被浪费的。

　　做这一幅《路》所剩下的三角形布头，我做了前面展示的作品《节俭之美》——很多布头会在库里安然沉睡十几个年头。但是不要着急，它们会不定时地赋予人灵感，更会于不经意中在某幅作品上发出魅力之光。

　　就像我在《节俭之美》中说的，真心希望大家一起收集布头，将它们从垃圾桶中拯救出来，为环保低碳做出贡献，也为自己的人生和创作做出贡献！

(局部 Details)

Road I, Wall-hanging

Silk yarn, 180cm×160cm, Back triple stitch, 1999

This *Road* is a road of my life that I can't describe in words. It is also the road to environment-friendly and low-carbon creation that I pursue.

Because my mother is the tailor, the various colorful cloths are my childhood toys. Until now, every time I see them, I still feel a warm feeling.

The material used in this work is the cloth cut when making traditional skirts of the Korean nationality. They are very long, just like a road. People will experience too much in their lives, and different feelings cannot be expressed in words. Sometimes, it will be frustrated like a cloth, being thrown into a trash can. Fortunately, my character is very tenacious and I will never give up.

After the completion of this work, I am immersed in the joy of happiness — it is really hard work! My life is spread out in front of this form. It wakes up my bit of memory like a movie... I believe that the unparalleled colors of the pieces that are spelled out by a piece of cloth will give each person the same beautiful feelings.

It is also the road to environment-friendly and low-carbon creation that I pursue. The reason why our limited life creates infinite value is because we use every bit of love, enthusiasm and time, without wasting any little touch and spirituality, just like using a little bit of cloth. Just like a piece of work that brings happiness to everyone, every little piece of cloth can play a major role in the whole work — they should also be cherished and not to be wasted.

The triangle cloths left by this work, I used them to make the work *Beauty of Frugality* that was introduced earlier — many cloths will sleep peacefully in the closet for more than ten years, but don't worry, they will give people inspiration from time to time, more inadvertently, the light of charm will be emitted on a certain work.

As I said in the *Beauty of Frugality*, I really hope that everyone will collect the leftover of cloths, save them from the trash can, contribute to the environmental protection and low carbon, and also contribute to their own life and creation!

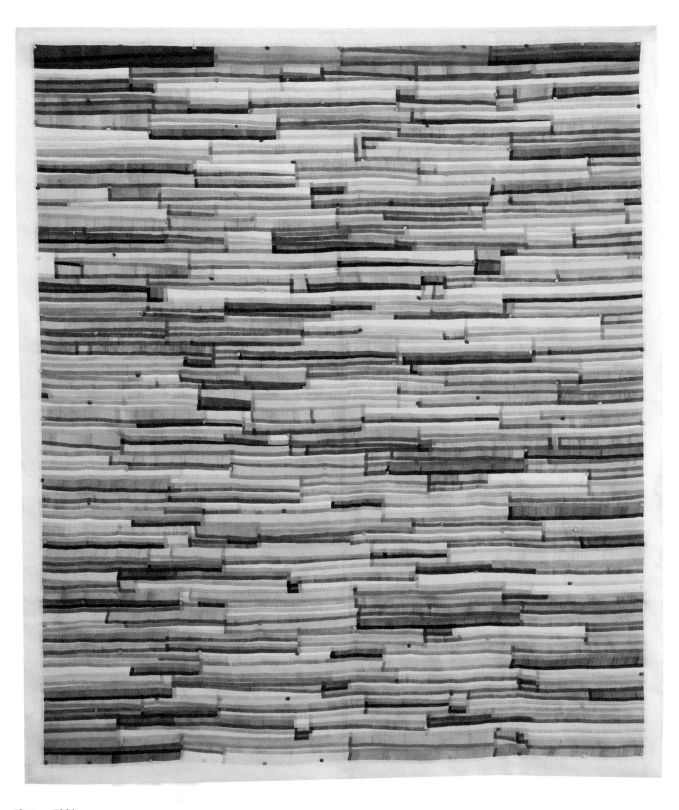

路 II　壁挂

真丝纱，190cm×165cm，植物染、倒三针，2004 年

　　我始终尽力使拼布图案减至单纯，没有多余的"废话"，不用过多的语言解释拼布的意义，也不揭示某种隐含的意义，只是通过呈现拼布的色彩来表达我心灵本身，让观者在我的拼布作品里发现其中层层隐现的谜。

Road Ⅱ, Wall-hanging

Silk yarn, 190cm×165cm, Plant dyeing & Back triple stitch, 2004

I always try my best to make the patchwork pattern simple, there is no extra nonsense, no need to explain the meaning of the patchwork with too much words, and no hidden meaning for revealing. I just use the color of the patchwork to express my heart. Let the audience themselves find the mystery on my patchwork.

(局部 Details)

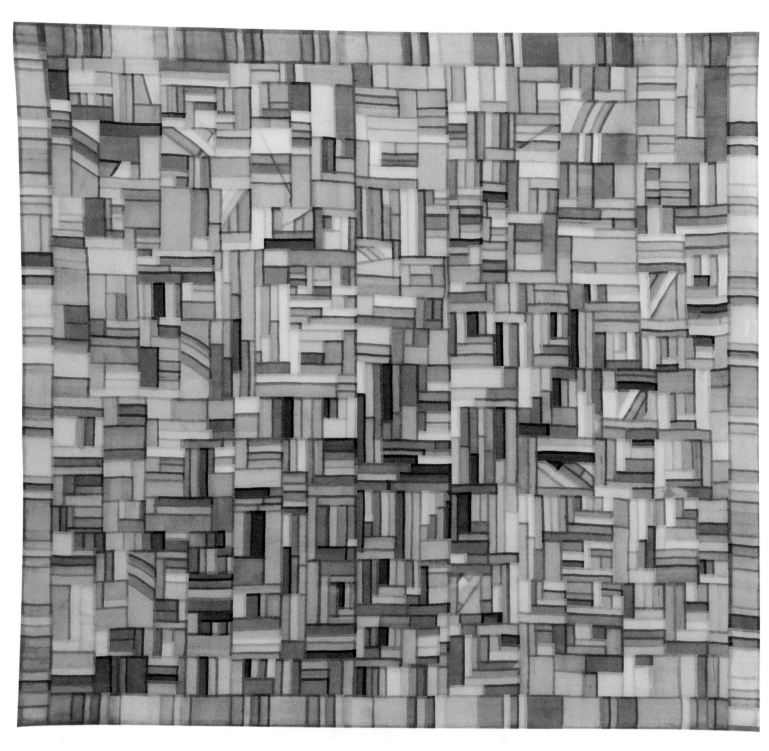

自然之美　壁挂

纱，200cm×180cm，植物染、倒三针，2010 年

　　该作品为应邀参加2012中国台湾国际拼布艺术展而专门制作。运用从各种天然植物中萃取的染料进行纯手工染色。在我的心目中，宝岛台湾是一个绿色之岛，环境优美、生态良好，所以选择用绿色调的布头拼成了台湾岛的形状，位于画面的中央。右下角的色调呈现优雅的粉色，那是我对香港的印象——时髦而雅致。左上角的部分则是祖国大陆，众多民族生活在一起，构成了多元的民族文化，再加上丰富的物产，所以呈现出色彩斑斓的美好景象。

Beauty of Nature, Wall-hanging

Gauze, 200cm×180cm, Plant dyeing & Back triple stitch, 2010

The work was specially made for the 2012 China Taiwan International Patchwork Exhibition. It is hand-dyed, using dyes extracted from various natural plants. In my mind, Taiwan is a green island with a beautiful environment. Therefore, the green cloth is used to make the shape of the island of Taiwan, which is located in the center of the picture. The shade in the lower right corner are pink, which is my impression of Hong Kong — stylish and elegant. The upper left corner is the mainland of China, where many ethnic groups live together, forming a diverse national culture, with rich products, it presents a colorful and beautiful scene.

(局部 Details)

壁挂

暗花罗，140cm×140cm，撩针，1994 年

Wall-hanging

A veiled design gauze, 140cm×140cm, Slip stitch, 1994

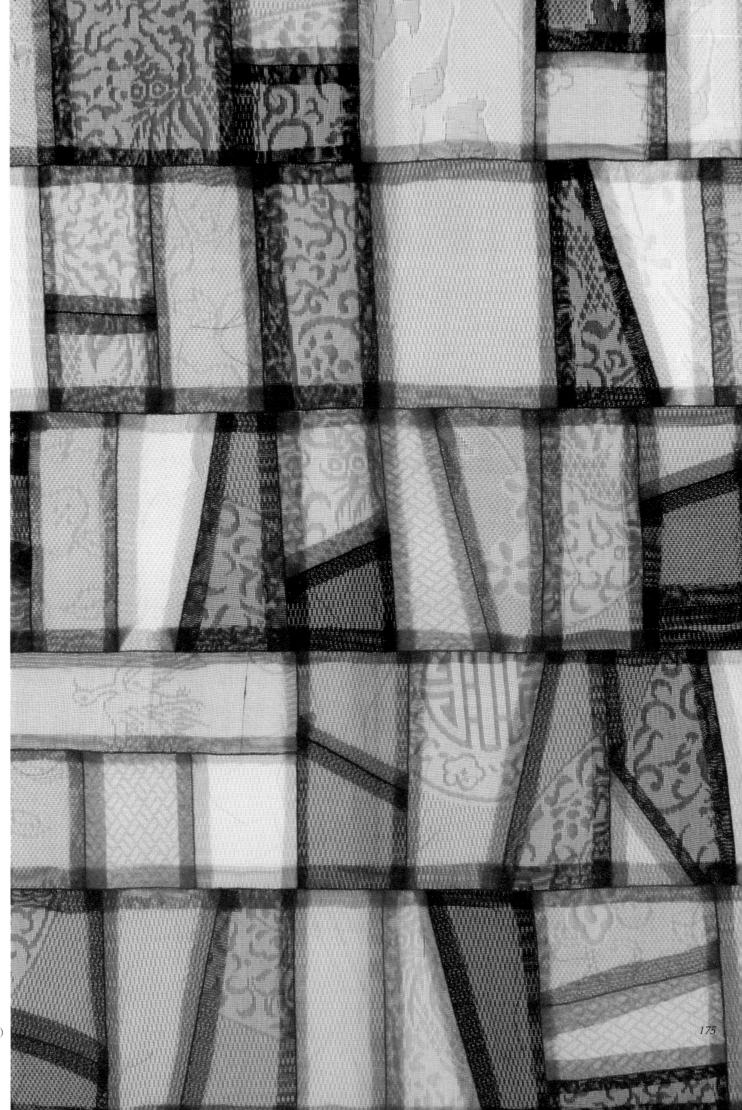

(局部 Details)

壁挂

纱，128cm×128cm，撩针，1992 年

按照布头的大小而随意拼成。

Wall-hanging

Gauze, 128cm×128cm, Slip stitch, 1992

Randomly assembled according to the size of the odd bits of cloth.

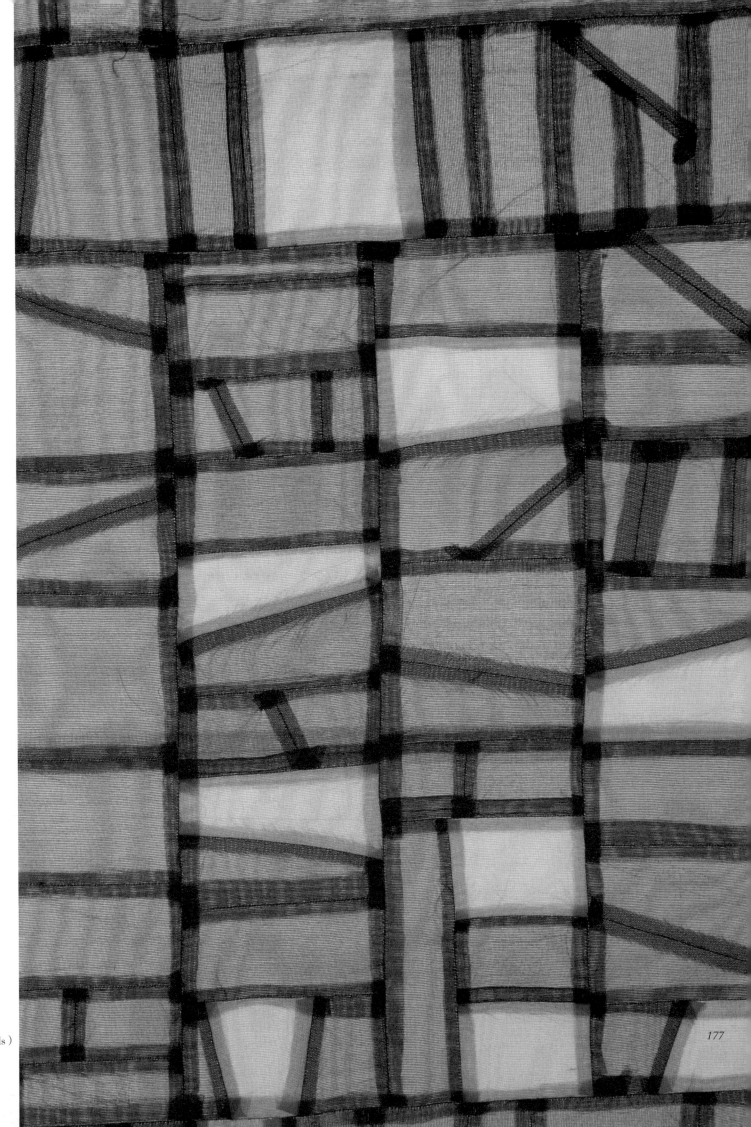
(局部 Details)

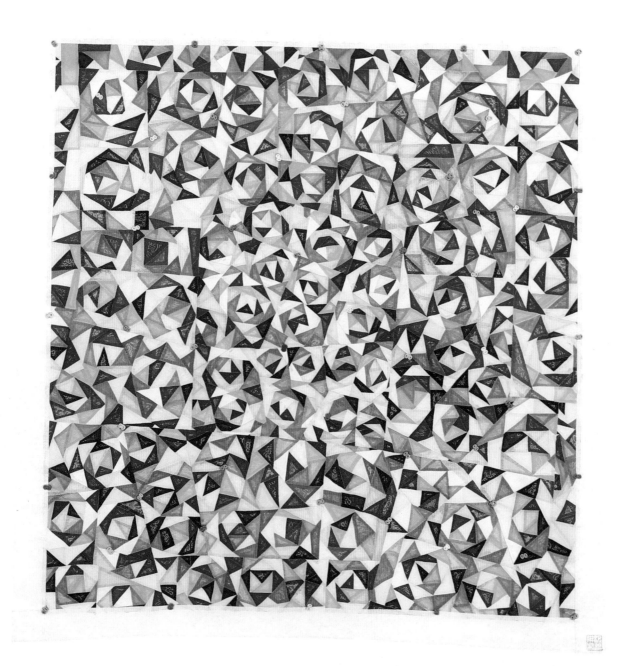

玫瑰花园　壁挂

暗花纱，108cm×108cm，撩针，1995年

　　这幅《玫瑰花园》是我对各种不同的三角形布头进行拼接的练习作之一。记得当时看着被丢弃在垃圾桶中的三角形五彩布头们，我的心里特别难受。怎么办呢？我当即决定：就用它们练习三角形的拼接吧。

　　用缭针拼接大小不同、形状各异的三角形布头是很有难度的，我就耐下心来慢慢做。在一块一块悉心拼接的时候，我突然发现了花的形状，当时特别兴奋：它们多像一朵朵盛开的玫瑰啊！

　　不禁感叹道：正是拼布这门手艺赋予了散碎布头们新的生命力，使它们得以再度回归生活、装点生活。在一块一块拼接、一针一针缝制的过程中，我体悟到：快乐幸福的拼布生活是需要智慧的——这也是手艺的智慧，物尽其用，变废为宝。

　　由这一幅《玫瑰花园》，我深深感到"花园"系列作品是充满幸福和智慧的，要继续下去！目前正在制作更加完美的《牡丹花园》作品。

(局部 Details)

Rose Garden, Wall-hanging

Veiled design gauze, 108cm×108cm, Slip stitch, 1995

  This *Rose Garden* is one of my exercises for splicing different triangle cloths. When I saw those colorful cloths dumped in the trash can, I was very upset. What should I do? I made my mind immediately that I would do some exercises of splicing triangle with them.

  It is difficult to splice those different-shaped triangle cloths, so I calm myself down and do it slowly. When I carefully stitched them together, I suddenly found the shape of the flower, which was particularly exciting at the time - they look like rose blossoms!

  I was very excited and overwhelmed. It is the patchwork technique that gives the rags a new vitality, allowing them to return to life and decorate their lives. During the process of making stitches and patches, I understand that the life of patchwork requires wisdom - this is also the wisdom of craftsmanship, making the best use of it, turning waste into treasure.

  From this work *Rose Garden*, I deeply feel that the "Garden" series is full of happiness and wisdom, and we will continue to create! Now I am making a more perfect Work — *Peony Garden*.

四、包袱皮
Bojagi

包袱皮（Bojagi）是朝鲜族最典型的拼布应用。其本来用途是为了包裹东西方便运输或遮盖东西防止灰尘。随着生活方式的变迁，包袱皮在今天更多地体现出朝鲜族注重礼节的人生准则和追求精致的生活情趣。这里的包袱皮大小不一，大的包裹被褥、服装，小的包裹礼金、饰物。都是用碎布头制作而成，但在拼接图案上体现出很大的差异，有非常规整的方形、三角形，也有较为随意的不规则图形。其中钱形纹的包袱皮则独树一帜，由于其图案具有美好的寓意，而被广泛运用于各种礼仪场合。

Bojagi is the most typical patchwork of Koreans. It was originally used to wrap things for transportation or covering things to prevent dust. With the change of lifestyle, the Bojagi today reflects the Korean people's etiquette and the pursuit of exquisite life. The Bojagi here vary in size, large one used for wrapping beddings and costumes, small one used for wrapping gifts and ornaments, all made out of oddments, but they show great differences in the stitching pattern. There are unconventional squares, triangles, and more random irregular patterns. The Bojagi with the money pattern is unique, and it is widely used in various ceremonial occasions because of its beautiful design.

惜福百世　包袱皮

纱，100cm×100cm（带长95cm），平针、握手缝，1990年

这幅《惜福百世》是复制之作，复制的对象是一件几百年前的民间老包袱皮。

1988年，我在首尔"丝田"刺绣博物馆看到了它——当我与那件已经有500年历史的老包袱皮面对面的时候，亲切感油然而生。记忆中，从前朝鲜族家家户户都自己做包袱皮。中国人是善以谐音寓吉的，所以"包袱"即"包福"，是一种很温暖的日常用品。

人们要懂得惜福，福气才会绵延下去，世世代代美满幸福。基于这样的愿望，在许东华馆长的同意下，我复制了这件作品，并为它起名《惜福百世》。

此后，我用各种面料来做这种包袱皮——也是从这时起，我将平针握手缝的技法应用到了拼布作品和拼布服装上。

(局部 Details)

Hundreds Years of Cherish, Bojagi

Gauze, 100cm×100cm (Belt 95cm), Plain stitches & Flat felled seam, 1990

  This *Hundreds Years of Cherish* is a copy, from a piece of traditional bojagi aged hundreds years old.

  In 1988, I saw it in the embroidery museum "Silk Farm" in Seoul; When I saw face to face with this 500 years old bojagi, a sense of intimacy raised from my heart. In my memory, being Korean nationality, every family used to make bojagi. Chinese people are good at homonym, a "bojagi" sounds like "wrap the fortune" in Chinese, so it is seen as a fortunate object.

  People should cherish their fortune; the fortune would last generation after generation. Based on this good meaning, under the permission of Director Xu Donghua, I made a copy of this work, and named it *Hundreds Years of Cherish*.

  From then on, I used all kinds of cloths to make bojagis. And from that time, I applied the technique of plain stitches into patchwork and patchwork garment.

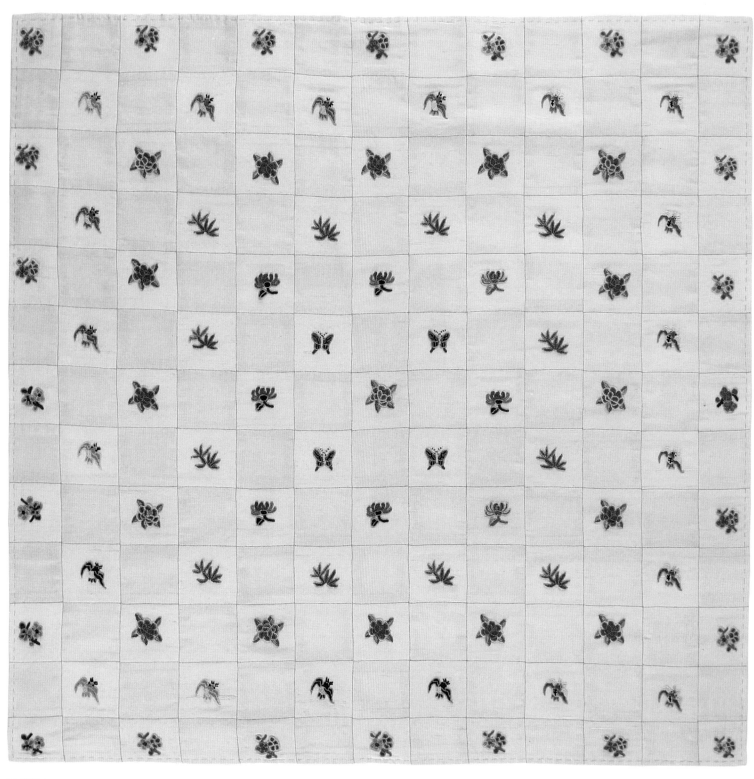

包袱皮

绸，66cm×66cm，撩针、打籽绣，1988 年

Bojagi

Silk, 66cm×66cm, Slip stitch & Embroidery, 1988

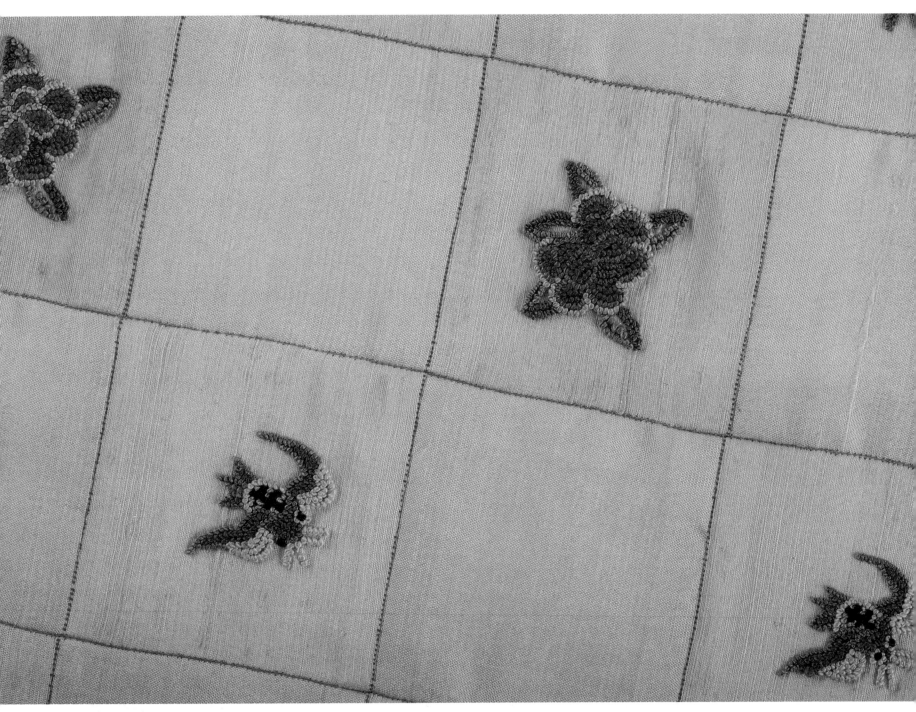

(局部 Details)

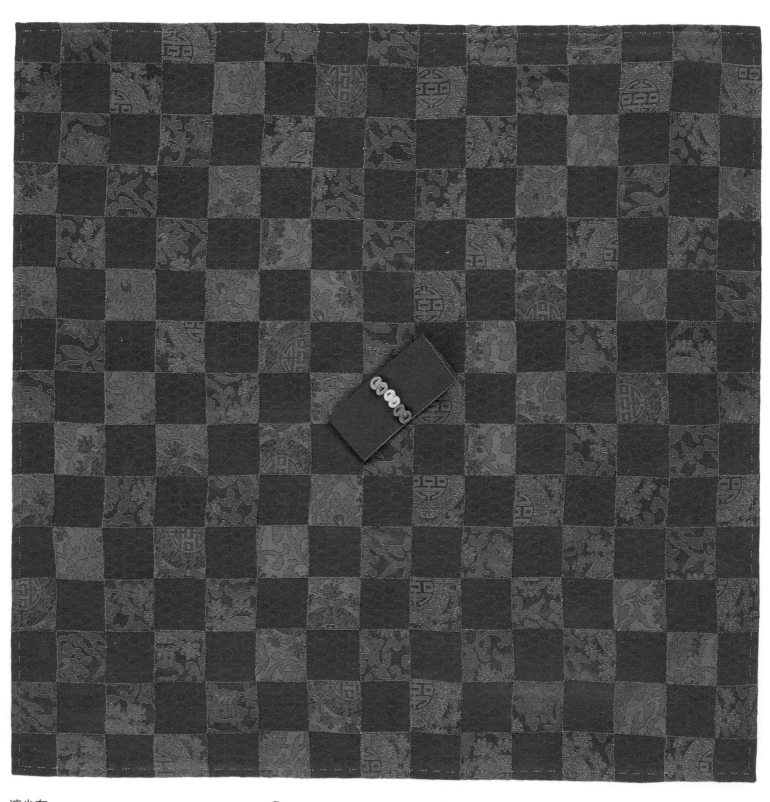

遮尘布

暗花缎，42cm×42cm，Slip stitch，1988 年

Cover

Veiled design satin, 42cm×42cm, Slip stitch, 1988

包袱皮

织锦缎，82cm×82cm，撩针，2003 年

Bojagi

Brocade, 82cm×82cm, Slip stitch, 2003

191

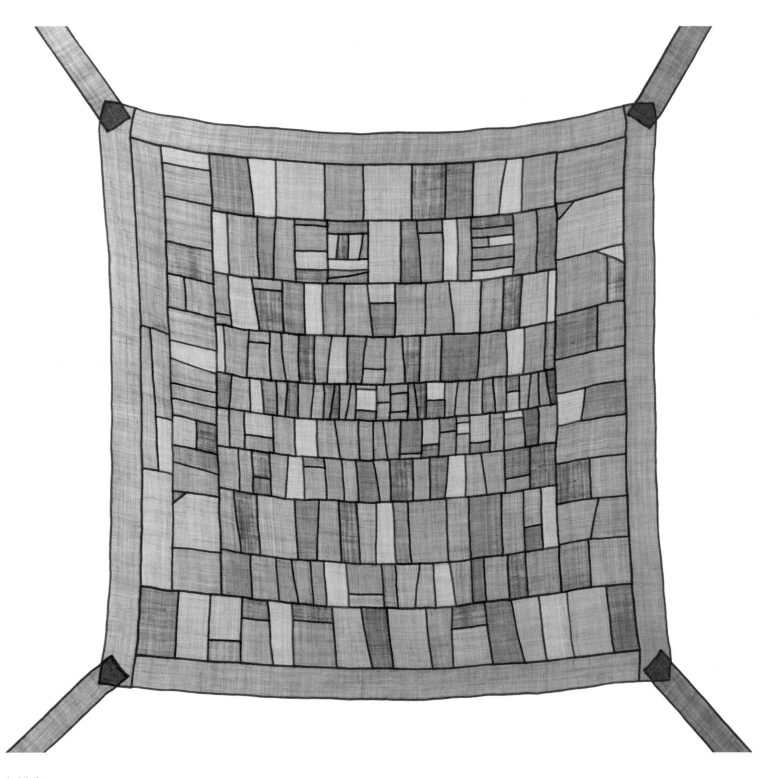

包袱皮

苎麻,95cm×95cm(带长120cm),植物染(蓝靛)、撩针、握手缝,1996 年

Bojagi

Ramie, 95cm×95cm (Belt 120cm), Plant dyeing(Indigo) & Slip stitch & Flat felled seam, 1996

(局部 Details)

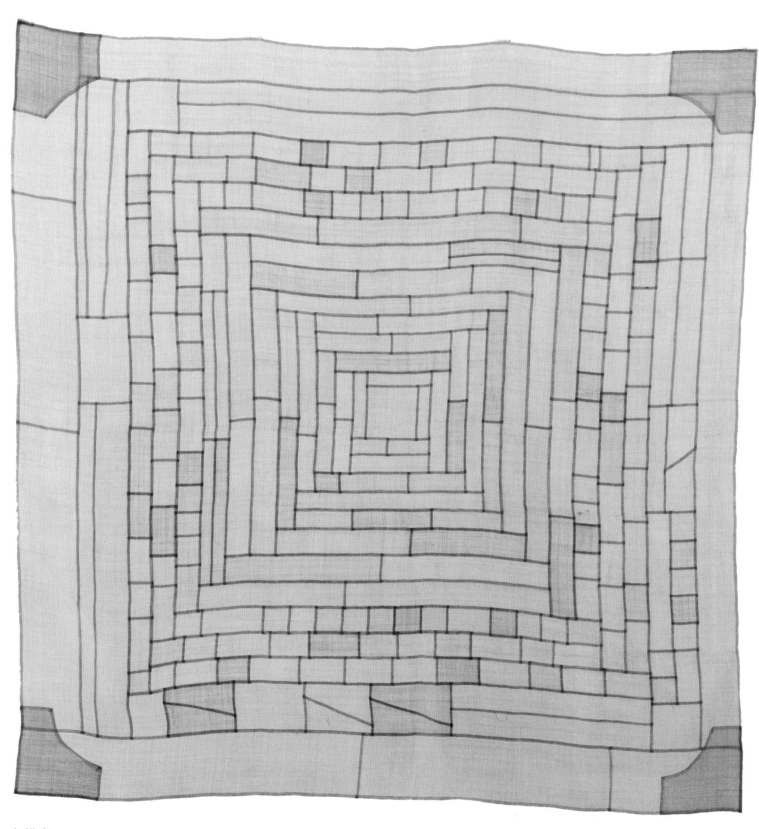

包袱皮

苎麻，88cm×88cm，撩针、握手缝，1988 年

Bojagi

Ramie, 88cm×88cm, Slip stitch & Flat felled seam, 1988

包袱皮

苎麻，50cm×50cm，撩针、握手缝，1995 年

Bojagi

Ramie, 50cm×50cm, Slip stitch & Flat felled seam, 1995

遮尘布

苎麻，40cm×40cm，撩针、刺绣、握手缝，1995 年

Cover

Ramie, 40cm×40cm, Slip stitch & Embroidery & Flat felled seam, 1995

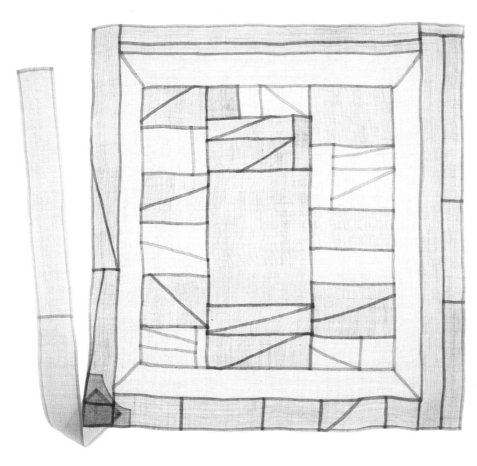

包袱皮

苎麻，50cm×50cm（带长 60cm），撩针、握手缝，1995 年

Bojagi

Ramie, 50cm×50cm (Belt 60cm), Slip stitch & Flat felled seam, 1995

包袱皮

苎麻,50cm×50cm(带长62cm),撩针、握手缝,1996年

Bojagi

Ramie, 50cm×50cm (Belt 62cm), Slip stitch & Flat felled seam, 1996

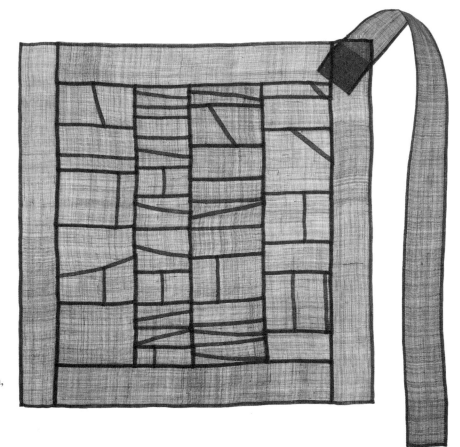

包袱皮

苎麻,50cm×50cm(带长62cm),撩针、握手缝,1996年

Bojagi

Ramie, 50cm×50cm (Belt 62cm), Slip stitch & Flat felled seam, 1996

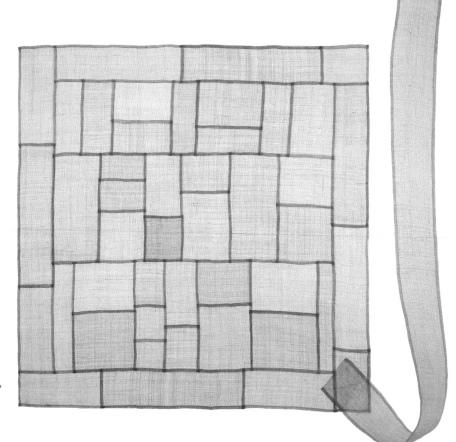

包袱皮

纱，88cm×88cm，倒三针，1998 年

Bojagi

Gauze, 88cm×88cm, Back triple stitch, 1998

包袱皮

纱，40cm×40cm，撩针，1992 年

Bojagi

Gauze, 40cm×40cm, Slip stitch, 1992

包袱皮

纱，37cm×37cm，撩针，1998 年

Bojagi

Gauze, 37cm×37cm, Slip stitch, 1998

包袱皮

纱，52cm×52cm，倒三针，1993 年

Bojagi

Gauze, 52cm×52cm, Back triple stitch, 1993

包袱皮

纱,45cm×45cm,撩针,1993 年

Bojagi

Gauze, 45cm×45cm, Slip stitch, 1993

遮尘布

纱,50cm×50cm,倒三针,1994 年

Cover

Gauze, 50cm×50cm, Back triple stitch, 1994

包袱皮

纱，51cm×51cm，倒三针，1990 年

Bojagi

Gauze, 51cm×51cm, Back triple stitch, 1990

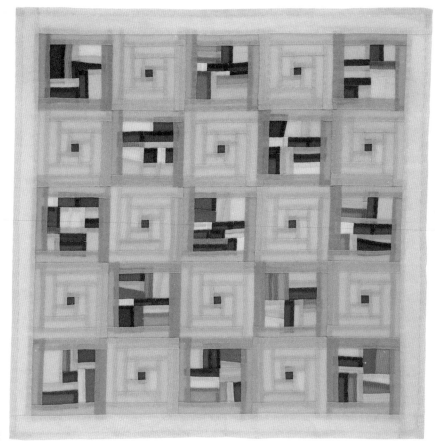

包袱皮

纱，57cm×57cm，倒三针，1992 年

Bojagi

Gauze, 57cm×57cm, Back triple stitch, 1992

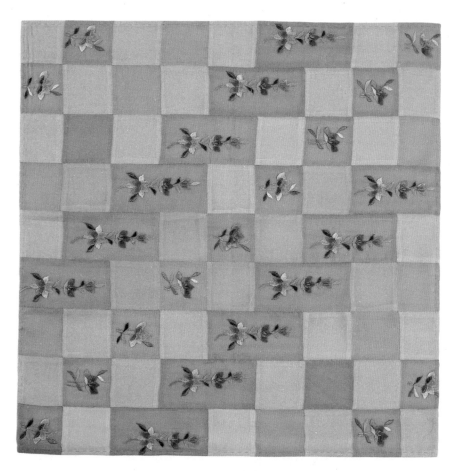

包袱皮

纱,53cm×53cm,机绣、撩针,1989 年

Bojagi

Gauze, 53cm×53cm, Machine embroidery & Slip stitch, 1989

包袱皮

暗花纱,37cm×37cm,Slip stitch,1991 年

Bojagi

Veiled design gauze, 37cm×37cm, Slip stitch, 1991

包袱皮

纱，57cm×57cm，撩针，1990 年

Bojagi

Gauze, 57cm×57cm, Slip stitch, 1990

包袱皮

纱，45cm×45cm，撩针，1989 年

Bojagi

Gauze, 45cm×45cm, Slip stitch, 1989

包袱皮

暗花纱，45cm×45cm，撩针，1990 年

Bojagi

Veiled design gauze, 45cm×45cm, Slip stitch, 1990

包袱皮

绸，75cm×75cm，撩针，1988 年

Bojagi

Silk, 75cm×75cm, Slip stitch, 1988

包袱皮

纱,46cm×46cm,撩针,1990 年

Bojagi

Gauze, 46cm×46cm, Slip stitch, 1990

包袱皮

暗花纱,47cm×47cm,Slip stitch,1980

Bojagi

Veiled design gauze, 47cm×47cm, Slip stitch, 1980

包袱皮

纱,50cm×50cm,倒三针,1996 年

Bojagi

Gauze, 50cm×50cm, Back triple stitch, 1996

遮尘布

纱,48cm×48cm,撩针,1989 年

Cover

Gauze, 48cm×48cm, Slip stitch, 1989

遮尘布

罗,46cm×46cm,撩针,1989 年

Cover

Leno, 46cm×46cm, Slip stitch, 1989

遮尘布

暗花纱,46cm×46cm,撩针,1989 年

Cover

Veiled design gauze, 46cm×46cm, Slip stitch, 1989

包袱皮

绸，49cm×49cm，倒三针，1993 年

Bojagi

Silk, 49cm×49cm, Back triple stitch, 1993

包袱皮

绸，50cm×50cm，撩针，1988 年

Bojagi

Silk, 50cm×50cm, Slip stitch, 1988

包袱皮

暗花纱，51cm×51cm，撩针，1989 年

Bojagi

Veiled design gauze, 51cm×51cm, Slip stitch, 1989

包袱皮

暗花纱，42cm×42cm，撩针，1989 年

Bojagi

Veiled design gauze, 42cm×42cm, Slip stitch, 1989

215

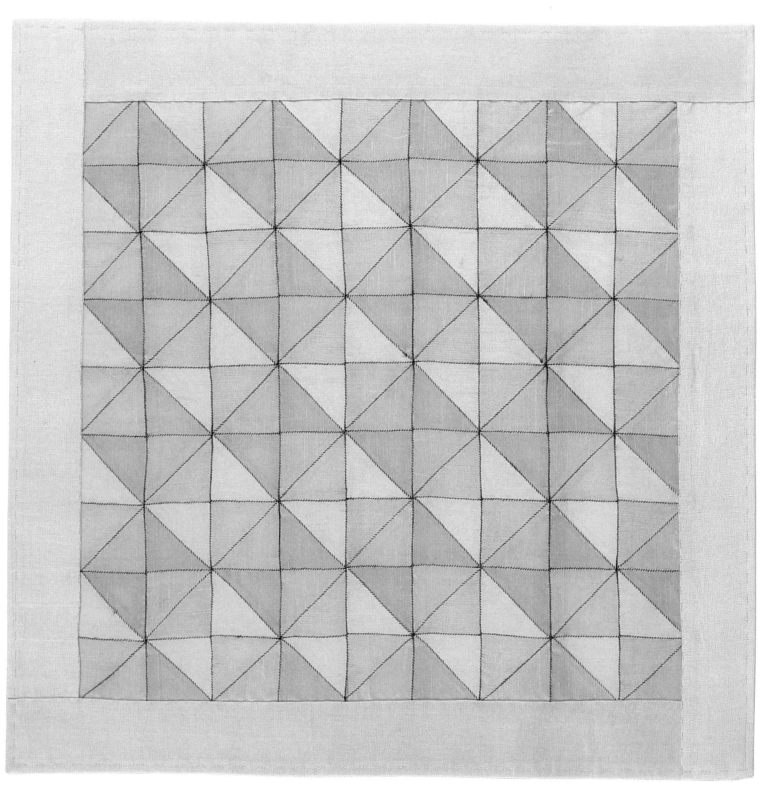

包袱皮

电力纺，39cm×39cm，撩针，1989 年

Bojagi

Habotai, 39cm×39cm, Slip stitch, 1989

包袱皮

电力纺，38cm×38cm，撩针，1989 年

Bojagi

Habotai, 38cm×38cm, Slip stitch, 1989

包袱皮

电力纺，32cm×32cm，撩针，1989 年

Bojagi

Habotai, 32cm×32cm, Slip stitch, 1989

包袱皮

暗花纱，39cm×39cm，撩针，1989 年

Bojagi

Veiled design gauze, 39cm×39cm, Slip stitch, 1989

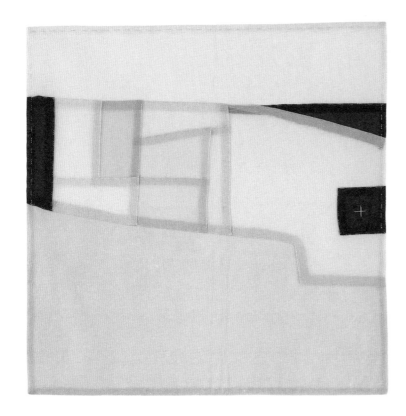

包袱皮

纱，36cm×36cm，倒三针，1988 年

Bojagi

Gauze, 36cm×36cm, Back triple stitch, 1988

包袱皮

暗花缎，33cm×33cm，撩针，1990 年

Bojagi

Veiled design satin, 33cm×33cm, Slip stitch, 1990

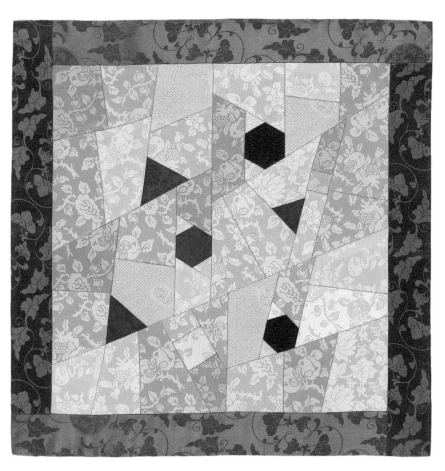

包袱皮

暗花缎，40cm×40cm，Slip stitch，1990 年

Bojagi

Veiled design satin, 40cm×40cm, Slip stitch, 1990

包袱皮

暗花纱，39cm×39cm，撩针，1991 年

三角形拼接作品的尝试。

Bojagi

Veiled design gauze, 39cm×39cm, Slip stitch, 1991

An attempt at triangular stitching.

温暖的味道　包袱皮

织锦缎，50cm×50cm，撩针，1998 年

《温暖的味道》——是母亲的记忆和味道。

这幅作品很特别，它所使用到的布头，均是母亲过去做服装时所剩下的仅有的丝绸老布头。拿出这些布头的时候，母亲讲了很多我童年的故事……

我将它们全部用到这幅作品中——唯有这一幅，在设计和制作的时候，我几乎没有改变这些老布头们的原始形状。

在我一针一线地制作时，会得到一种发自心灵深处的、最透彻的满足。因为在这些老布头中，我感受到了最温暖的味道。

不需要太多的语言，每一片老布头都化成我心中最美好的一片记忆。爱是最大的动力和指引——要继续前进，"因爱而做"是我人生最美的主题。

The Taste of Wormth, Bojagi

Brocade, 50cm×50cm, Slip stitch, 1998

The Taste of Warmth — the taste and memory of Mother

This work is special, the cloths are the leftovers or old silk from my mother's work. When I took them out, mother told me lot's of stories of my childhood...

I put all of them in this piece. Only this one, I barely changed the shape of the cloths during the process of design and making.

When I made this one, stitch after stitch, I felt the thorough sarisfaction deeply in my heart. Because in these old cloths, I feel the warmth of my mother.

No need of more words, every piece of old cloth would be the best memory in my heart. Love gives the best motive and direction - I will continuously move on. "Do it for love" is the best theme of my life.

225

包袱皮

纱、暗花缎，43cm×43cm，倒三针、翻折，1994 年

　　钱形纹是将折叠好的方形布块连接在一起后，再翻折形成一个一个外圆内方的铜钱相连的图案。相同的图案在欧美被称为"教堂之窗"；在韩国叫作如意纹，做法稍有不同。中国新疆库车博物馆陈列着一幅出土的唐代盖脸布也运用了钱形纹，九个一组。中国的钱形纹不仅有很早的出土实物例证，而且在民间广为流传，壮族、彝族、白族、满族均有运用。这是财富的象征，而且制作方法更加丰富。据我在少数民族地区所看到的就已经有七种做法，但最终呈现的纹样都是钱形纹。

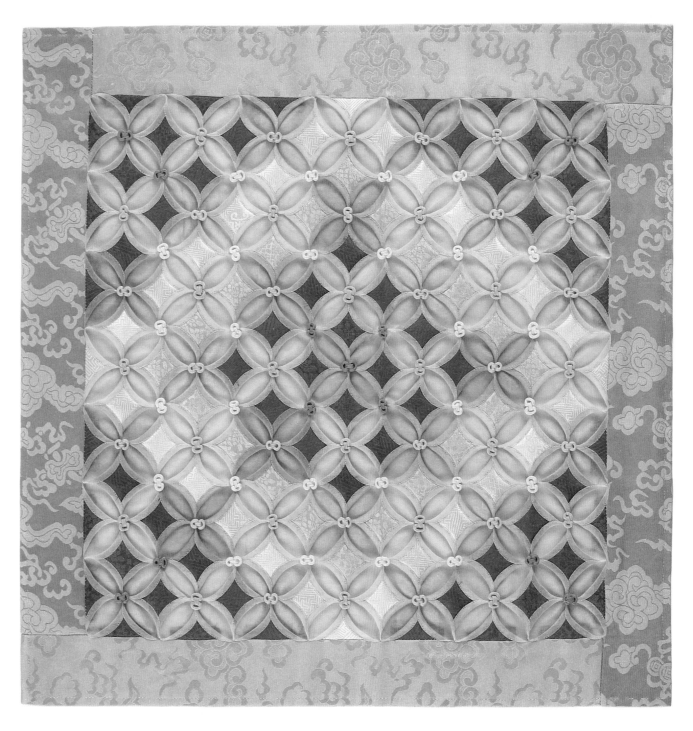

Bojagi

Gauze & Veiled design satin, 43cm×43cm, Back triple stitch & Folding, 1994

  The money pattern is a pattern in which the folded square pieces of cloth are joined together, and then turned over to form connected copper coins which are square internally and round externally. The same pattern is called "the window of the church" in Europe and America. In Korea, the money pattern is called the "Ruyi" (as-you-wish) pattern, which is slightly different in making. In the Kuqa Museum in Xinjiang, China, an unearthed face-covering cloth from the Tang Dynasty also shows a money pattern, nine for a set. The Chinese money pattern not only is proved by early unearthed articles, but also is widely popular among the folk.

包袱皮 **Bojagi**

纱、暗花缎，44cm×44cm，倒三针、翻折，1994 年 Gauze & Veiled design satin, 44cm×44cm, Back triple stitch & Folding, 1994

(局部 Details）

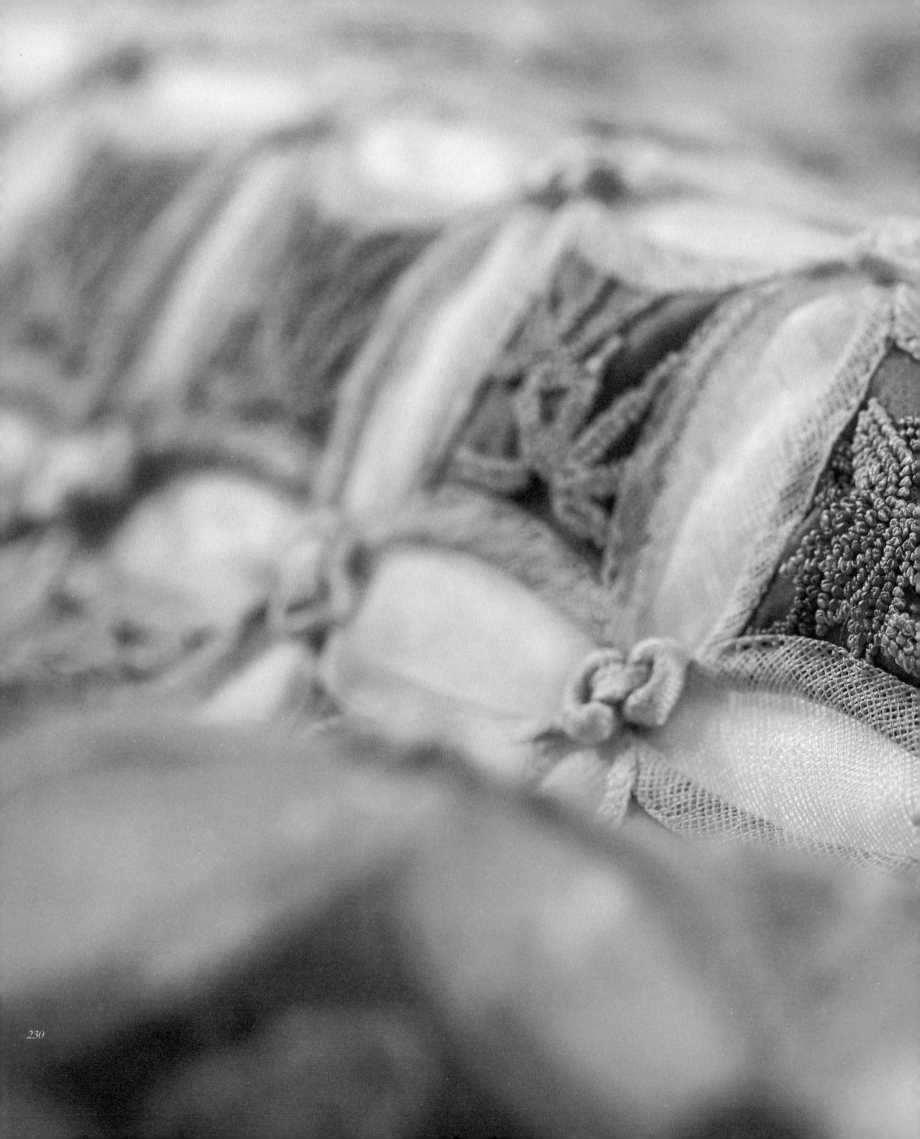

包袱皮

暗花纱、软缎，90cm×90cm，倒三针、翻折、打籽绣，2003 年

Bojagi

Veiled design gauze & Satin, 90cm×90cm, Back triple stitch & Folding & Dazi embroidery, 2003

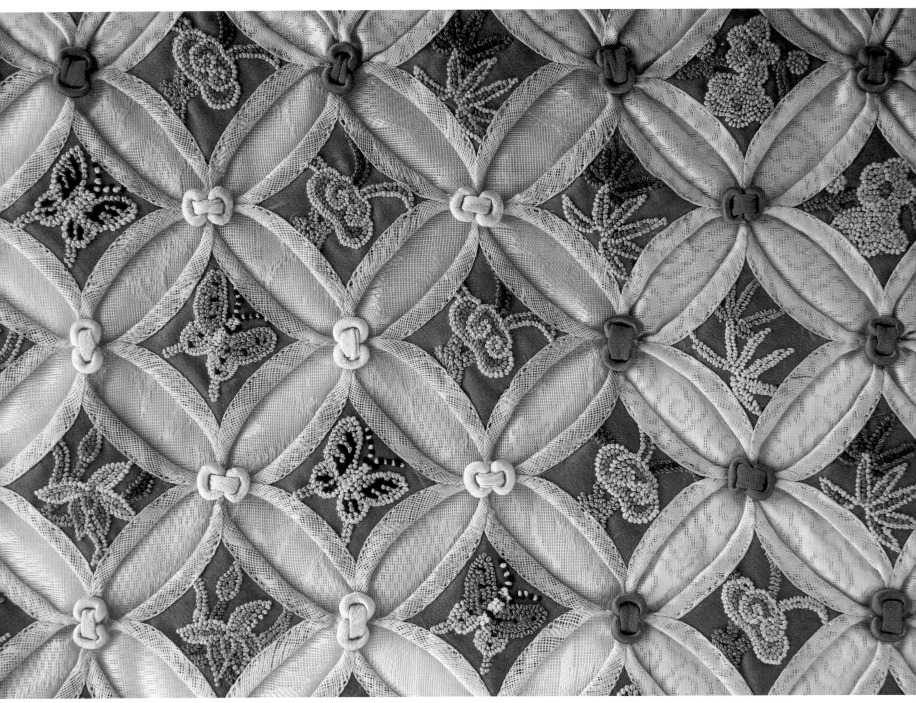

(局部 Details)

235

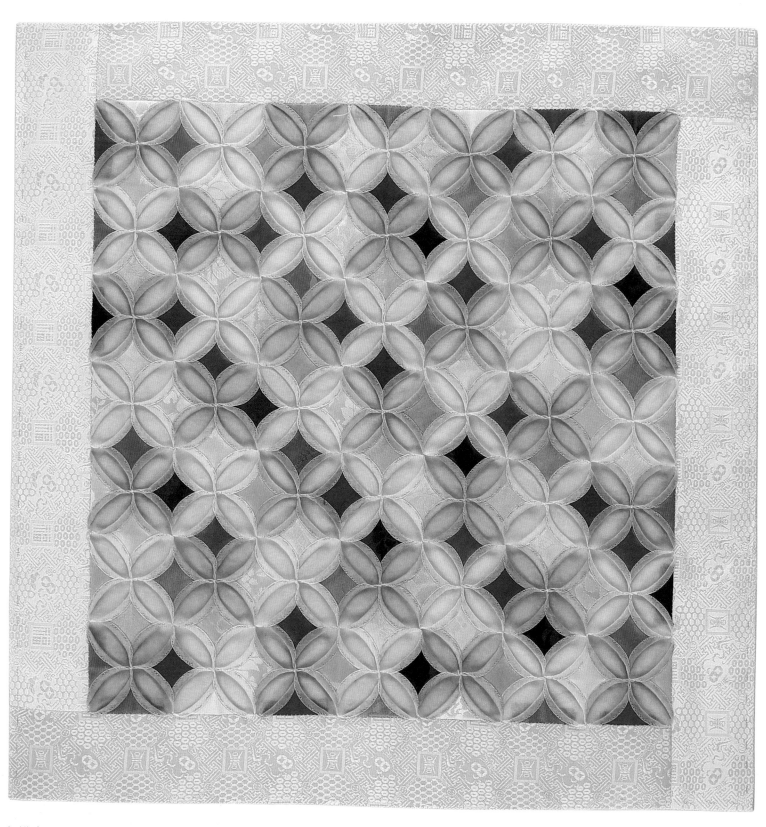

包袱皮

纱、暗花缎，38cm×38cm，倒三针、翻折，1994年

Bojagi

Gauze & Veiled design satin, 38cm×38cm, Back triple stitch & Folding, 1994

包袱皮

纱、暗花缎，37cm×37cm，倒三针、翻折，1994 年

Bojagi

Gauze & Veiled design satin, 37cm×37cm, Back triple stitch & Folding, 1994

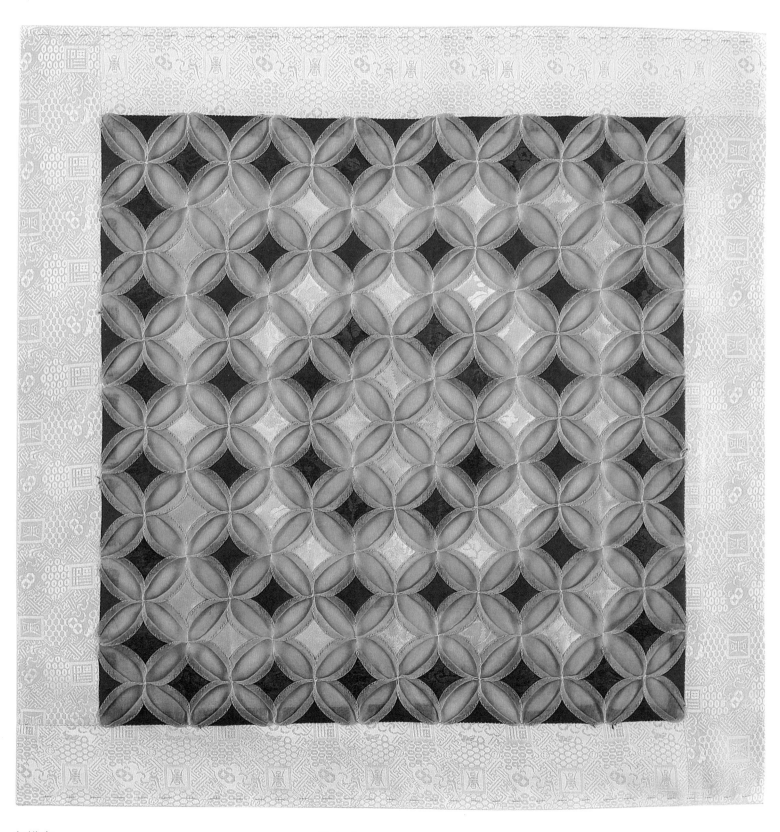

包袱皮

纱、暗花缎，38cm×38cm，倒三针、翻折，1994 年

Bojagi

Gauze & Veiled design satin, 38cm×38cm, Back triple stitch & Folding, 1994

(局部 Details)

包袱皮

纱、暗花缎，43cm×43cm，倒三针、翻折，1994 年

Bojagi

Gauze & Veiled design satin, 43cm×43cm, Back triple stitch & Folding, 1994

包袱皮

纱、软缎,40cm×40cm,倒三针、翻折,1994 年

Bojagi

Gauze & Veiled design satin, 40cm×40cm, Back triple stitch & Folding, 1994

包袱皮

纱、软缎，45cm×45cm，倒三针、翻折，1994 年

Bojagi

Gauze & Satin, 45cm×45cm, Back triple stitch & Folding, 1994

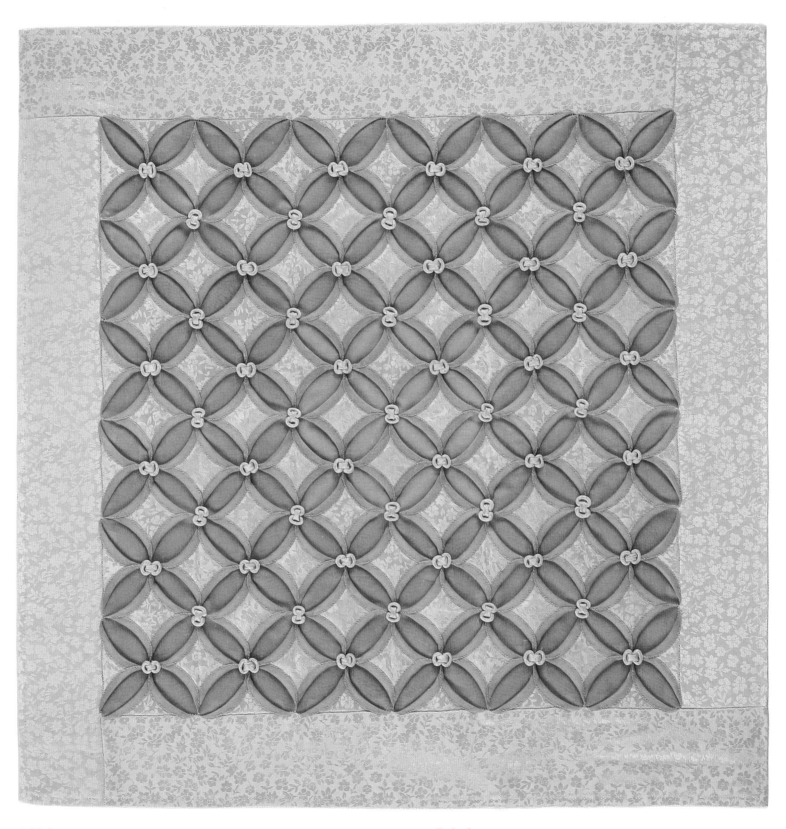

包袱皮

纱、暗花缎,43cm×43cm,倒三针、翻折、植物染(红花和蓝靛),1996年

Bojagi

Gauze & Veiled design satin, 43cm×43cm, Back triple stitch & Folding & Plant dyeing (Safflower and indigo), 1996

五、习作与片段

Exercises and Pieces

人们看到我的拼布成品往往会惊叹不已，而在这些成品背后，还有更多数量的试验品、半成品以及失败的作品，这些有时更能体现我的创作历程与艰辛。

People are often amazed at my finished patchworks. Actually, there are a great number of experimental works, semi-finished and failed works, which somewhat can better reflect my creative process and hardship.

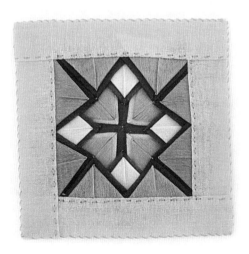

试验品

纱,叠花,1995 年

这是在我看到贵州施洞苗族的堆绣之后进行的尝试。首先我选择了常用的真丝绡,直接折叠而成。从一个单元,到一个长条组合,再到一个方形的组合,最终从模仿中得到启发,改变了创作思路,产生了作品《百花争艳》。

Experimental piece

Gauze, Folding, 1995

This is an attempt I have made after seeing the barbola of the Shidong Miao minority. First of all, I chose the silk gauze, which is common for me, and then folded it. From a unit, to a combination of strips, to a combination of squares, it was finally inspired by the imitation, after changing the creative ideas, the work *Hundreds Flowers Blossoming* was done.

半成品

纱，撩针，1998年

　　我的所有作品，从第一步、第一针开始就是全神贯注的。我会用最大的耐心、不惜繁复工序地对待每一块布头和每一个步骤：下针、修剪、停止、观察、思考、再下针……这是一个循环，贯穿整个制作过程。

　　我做一件作品，首先在脑海中定好主题，按照主题构思色彩，然后动手做小块的拼接，最后再将一个个小块连成整张大幅作品。尤其在最后一步——将一块块拼好的拼布小块再放在一起拼排比对的过程中，会随时发现不足，然后反复补足、反复调整，无限还原拼布的平面感，一点点靠近自己所想要的效果……最后，再运用各种技法，细致入微地拼缝连接起来。

　　我的每一点用心都凝结在一针一线中。在这个过程中，我的内心是平和并且充满欢喜的——每每完成一件精致的作品，感受着自己的细致拼缝所赋予每一块布头的新的生命力，我都会心跳、沉醉，陷入到不愿醒来的、梦幻般的幸福中。

　　对我来说，拼布是挚爱、是生活中不可或缺的如呼吸一样自然而然的存在、是我的生命。每一件作品，都寄托着我对生活的热情与心中的沉静和思考……

Semi-finished products

Gauze, Slip stitch, 1998

    All my works, from the first step and the first stitch, I put all my effort and attention to it. I will treat each piece of cloth and every step with the utmost patience and indulgence. Stitch, trim, stop, observe, think, and then stitch again… This is a loop that fills the entire production process.

    When I work, I will set the theme in my mind and conceive the color according to the theme firstly, and then make small pieces of stitching. Finally, I will connect the small pieces into a whole large piece. Especially in the last step - in the process of putting pieces of patchwork together, the defects will be found at any time, and then I repeat and adjust, infinitely restore the flatness of the patchwork, a little bit close to the effect I want... At last, I use a variety of techniques to splicing and joining together.

    Every single point of my heart is condensed in my stitches and strings. In the process, my heart is calm and full of joy — every time I finish a piece of work, I feel the delicate seams I have given to each piece of cloth. With new vitality, I feel the heartbeat, and I am intoxicated, and I fall into the dreamy happiness that I don't want to wake up.

    For me, patchwork is my love, an indispensable part of life, a natural existence like breathing — it is my life. Every piece of work has borne my passion and thought of life...

习作

绸,倒三针,1992 年

在观赏传统服饰以及走访少数民族地区的时候,我看到倒一针、倒二针、倒四针、倒五针、倒六针的作品,针脚细致、制作精美。回来之后我尝试了第一幅倒三针作品,针脚极尽所能的精细,真的像人们形容的那样"细如鱼籽",所谓"一芝麻三针"。传统服饰和少数民族服饰上的倒三针都是为了增加美感和加固而做的。但我在制作的过程中,感觉倒三针大有学问。其步骤是:倒一针,倒一针,再倒一针,总共往后退了三针,然后往前走一针,最后这一针比倒的三针的距离总和还大一点。所以说,经过退三步反复思考后再大跨步向前迈出去,即"三思而后行"。这恰恰体现了我们中华民族经反复思考后稳稳当当地往前走的民族风格。从那之后我几乎把倒三针运用在了我所有的拼布作品中,为的就是把这种民族品格注入我的一针一线之中。

Exercise

Silk, Back triple stitch, 1992

When watching traditional costumes and visiting ethnic minority areas, I saw the patchworks of reverse one stitch, two stitches, four stitches, and five stitches. The stitches were meticulous and exquisite. After I came back, I tried the first work of back triple stitch. The stitches are as fine as they can be, as people describe as "thin as fish roe", and the so-called " Three stitches in a sesame". The back triple stitch on traditional costumes and ethnic costumes are made to add beauty and reinforcement. But in the process of making it, I feel there is an art in the back triple stitch. It is one stitch back, one more stitch back, one more stitch back, a total of three stitches back, and then one stitch forward, the stitch forward is a little larger than the three stitches back. Therefore, after three steps of rethinking and then striding out, that is, "think twice and then go". This is precisely the style that the Chinese nation has gone through in the past after repeated thoughts. Since then, I have applied almost back triple stitch to all my patchwork, in order to inject this national character into my needlework.

设计稿

纱，粘贴，2017年

之前我做拼布都没有所谓的设计稿，整体的样子都在我的脑子里，而且一边制作，我会一边调整。只有一些复制的作品，我会事先做一个小稿。近些年为了方便给学生示意，我也开始尝试制作设计稿，是用糨糊把面料粘在纸上。

但这个设计稿不是给学生看的，而是给我的儿媳妇准备的。我对她说：我心中最可亲可爱的大儿媳莲玉，这个作品是送给你的礼物，叫作《生命的喜悦》。我把这个作品复制四份拼到一起时看到了无数的笑容和无数的小鱼。我深深感受到人和人之间真的是有着心灵的交流和感应！当我看到此作品时发现生命的平安与喜悦已经悄悄进入我的内心，我也希望当你看到此作品时同样能感受到内心的平安与喜悦。所以我把作品的名字定为《生命的喜悦》，希望你在人生路上一路修行，一路平静。你要相信，只要你自己决心要改变心中的世界，世界就会改变的！你也会惊讶于奇迹真的无所不在！因为你的努力，不仅改变了自己，同时也帮助了别人！每当我感觉内心不平静的时候，我认为就是我向内看的机会。这一路上我根本不担心，因为我愿意改变，我内在的智慧会引导我如何反省、如何去爱与被爱。

我感觉很有力量，很有勇气。我不会迷失，也从不孤单，因为爱与光明一直引导我走在正确的路上。在我人生路上，家人和拼布始终给我带来笑容和希望。你也跟我说过："我是老韩家大儿媳妇，我肯定能战胜一切！"希望你的生活和我作品的名字一样，一切都喜悦。感恩有你，我最爱的儿媳。

儿媳妇回复我：我最敬爱的妈妈，首先我要感谢您对我的关心和爱！我始终感恩妈妈您在我身边鼓励和支持！您每天的问候给了我最大的安慰！从看到《生命的喜悦》——妈妈您特意给我制作的作品时，我都能感受到浓浓的温暖。我不知多少次说过，妈妈您是我人生路上的榜样！您始终是用行动告诉我，人要为自己的梦想而加油努力，不管遇到什么困难都要坚持。您也常用话语开导我，经历挫折和磨难其实就是为了让自己内心更加强大而做准备。您鼓励我坚强勇敢地面对生活给予我的一切，尽管它并不是很完美也要微笑着面对！我会珍惜您给予我的一切教导并努力做好，不会辜负您对我的期望。我要为了成为您的骄傲而更加坚定信念，绝不会让丝毫负能量影响我的情绪，而是正能量满满地笑迎每一天！妈妈放心吧！我一定会加油的！妈妈，我爱您！

Design sample

Gauze, Pasting, 2017

Before I did the patchwork, there was no so-called design sample. The whole appearance was in my mind, and I would adjust it as I make it. There are only a few copied works, I will make a small draft. In recent years, I try to make a design draft for the convenience of students, which is to paste the fabric on the paper with paste.

This design sample is not for the students, but for my daughter-in-law. I said to her: The most dear and lovely child in my heart, Lian Yu, this work is a gift for you, called *The Joy of Life*. When I put together these four pieces, I saw countless smiles and countless small fish. I deeply feel that there really exist the communication and interaction of the soul! When I saw this work, the peace and joy of life quietly entered my heart. I hope that when you see this work, you will feel the peace of mind. So I called the name of the work *The Joy of Life*. I hope that you will practice all the way in your life, and you will be calm all the way. You should believe that as long as you are determined to change the world in your heart, the world will change! You will also be surprised that the miracle is really ubiquitous! Because of your efforts, not only have you changed yourself and your own world, but also helped your family and others! Whenever I feel that my heart is not calm, I think it is the opportunity for me to look inside. I am not worried about it at all, because I am willing to change, my inner wisdom will guide me how to reflect, how to love and be loved.

I feel powerful and courageous. I will never be lost, never alone, because love and light have always guided me on the right path. My family and patchwork always bring me smiles and hopes. You also told me, "I am the daughter-in-law of the oldest son of Han family, I am sure to conquer all difficulties in my life!" I hope that you will be happy as the name of my work, and I am grateful to have my beloved daughter-in-law.

My daughter-in-law replied to me: My most beloved mother, first of all, I would like to thank my beloved mother for concern and love for me! I am always grateful for your encouragement and support! Your daily greetings gave me the greatest comfort! I can feel the warmth from your work of *The Joy of Life* which made specially for me. You are the model on my life! You have always told me by action that one should work hard for his dream and persist regardless of any difficulties. Your words enlightens me to go through frustration and suffering. And makes my heart stronger. You encourage me to be brave and courageous to face everything that life gives me, even though it is not perfect, and I have to keep smile! I will cherish all the teachings you have given me and try my efforts to live up to your expectations. I want to be your pride. I will be more determined to never let negative energy affect my emotions. I will welcome each day by positive energy! Mom can rest assured! I will definitely try my best and insist on! Mom, I love you!

制作篇
Production

一、蝙蝠花的制作

Production of Batflower

二、太阳花的制作

Production of Sunflower

三、针线盒的故事

Story of A Sewing Box

一、蝙蝠花的制作

这是作品《百花齐放》的组成单元，同尖角花的组合方式一致，但是花瓣的制作不同。同样也是可以做成配饰或组合成大的作品。

一、Production of Batflower

This is one unit of the work *Hundreds Flowers Blossoming*, which is consistent with the combination of sharp-edged flowers, but the production of petals is different. It can also be made into accessories or combined into a large work.

将真丝绸面料剪成边长3cm的正方形。

Cut the silk fabric into a square with a side length of 3cm.

将上过浆的棉线剪成4cm长，放在小正方形内侧的一个角上向下卷。

Cut the starched cotton thread into 4cm and roll it down on a corner inside the small square.

卷到中央的位置，用珠针固定。注意卷紧，不要散开。

Roll it to the center and fix it with a bead needle. Pay attention to tighten it and do not spread it.

将卷好的布料对折，形成一个小圆弧。

Fold the rolled fabric in half to form a small arc.

用针线固定后，沿着线迹的边缘剪掉多余的部分，完成一个花瓣。

After fixing with needle and thread, cut off the excess part along the edge of the stitch to complete a petal.

6

准备一块底布，上面画出一个正圆形，大小根据需要而定。将上一步骤准备好的小花瓣缝合在底布上。接着固定第二个花瓣，紧紧挨着第一个花瓣，每个花瓣的间距相等，直到填满一圈。

Prepare a base fabric with a perfect circle drawn on it, which size is determined as need. Stitch the small petals prepared in the previous step on the base fabric. Then fix the second petal next to the first petal, and the spacing of each petal is equal until all these petals fill a circle.

7

完成第二圈，比第一圈圆形半径少0.5cm左右，但是花瓣的间距应该跟第一圈的间距相等。

Complete the second circle, which is about 0.5cm smaller than the circular radius of the first circle, But the spacing between petals should be the same as that of the first circle.

8

完成第三圈。注意每个花瓣顶部的反向延长线都要对准圆心，间距一致，呈现正圆形放射状的效果。

Complete the third circle. Note that the reverse extension line of the top of each petal should be aligned with the center of the circle, with the same spacing and a radial effect.

9

完成第四圈。每一圈的色彩由深到浅。

Complete the fourth circle. The color of each circle is from deep to shallow.

10

完成第五圈。越往后缝合的面料层次会越多，难度也会增加，要注意控制好力度。

Complete the fifth circle. The more layers of fabric, the more difficult it will be, and the more attention should be paid to control the strength.

11

第六圈是花蕊部分，这一层花瓣一定为6个，所以要严格调整花瓣位置的对称。

The sixth circle is the flower bud, which has fewer petals, Must be six. so the symmetry of the petals should be strictly adjusted.

12

最中心用一个蝙蝠结装饰，这样所有的缝线都隐藏在花瓣之下。

The center is decorated with a bat knot, so that all stitches are hidden under the petals.

二、太阳花的制作

这是作品《百花争艳》的组成单元，灵感来源于贵州施洞苗族独特的制作工艺——堆绣。用真丝的小方块折叠成小三角，再堆叠在一起组成图案。与作品《百花齐放》的折叠方式稍有不同，它运用各种颜色的真丝绡布头，组成一个个圆形的花朵单元。每一个花朵可以单独作为胸花、戒指、耳环等配饰，也可以组合成大幅的作品。

二、Production of Sunflower

This is the unit of the work *Hundreds Flowers Blossoming*, which is inspired by the unique craftsmanship of the Miao minority in Shidong, Guizhou – barbola. Fold the small triangles into small triangles and stack them together to form a pattern. The folding method of *Hundreds Flowers Blossoming* is slightly different. The silk fabric of various colors are used to form a circular flower unit. Each flower can be used alone as a boutonniere, ring, earrings, etc. Or it can be combined into a large piece.

1

将真丝绡面料剪成边长为3cm的正方形小方块，注意边缘须平行于面料的丝道。

Cut the silk gauze fabric into squares with a side length of 3cm, paying attention to the edges of the fabric parallel to the silk thread.

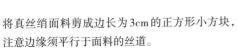

2

将小方块的一边向内折叠0.5cm。

Fold the side of the small square inward by 0.5cm.

3

沿着与折叠边垂直的一侧的中线，向内侧折叠一角，对齐中线。

Along the center line of the side perpendicular to the folded edge, fold a corner inward and align the centerline.

4

向中线折叠另外一角，形成一个90度的角。注意对折的时候不要拉扯面料，而是用拇指和食指轻轻按压，以防面料变形。

Fold the other corner to the centerline to form a 90 degree angle. Be careful not to pull the fabric when folding, use your thumb and forefinger to press gently that prevent the fabric from deforming.

5

沿着中线对折，形成一个45度的小三角。注意即使很薄的真丝绡也是有厚度的，所以上一步骤对齐中线的时候可以稍微留出1mm的空隙，这样折出的三角才是尖的。

Fold along the midline to form a small triangle of 45 degrees. Note that even a very thin silk gauze has a thickness, so the upper step can be slightly left with a 1 mm gap when the center line is aligned, so that the folded triangle is pointed.

6

制作花蕊部分的小三角，步骤同前。

Make a small triangle for the bud part, the same as before.

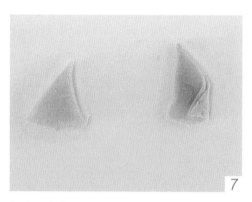

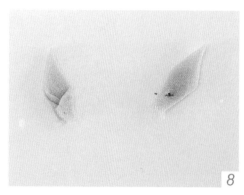

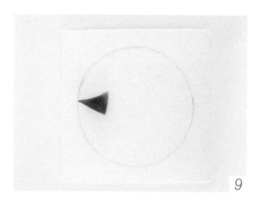

把小三角修剪成一个等腰三角形，底角向内折叠。

Trim the small triangle into an isosceles triangle, fold the bottom corner inward.

底角向内折叠，并用针线固定。将所有用于制作花瓣和花蕊的小三角都准备好，可以用针线串在一起备用。

Fold the bottom corners inward and secured with needlework. Prepare all the small triangles for making petals and stamens, and use needles to string together.

准备一块底布，上面画出一个正圆形，大小根据需要而定。将上一步骤准备好的小三角修剪成一个等腰三角形，然后尖部朝外，对角线对准圆心，用针线固定在底布上。

Prepare a base fabric with a perfect circle drawn on it, which size is determined as needed. Trim the small triangle prepared in the previous step into an isosceles triangle. The tip is facing outward, and the diagonal is aligned with the center of the circle. Then fix it on the base fabric by a needle.

接着固定第二个小三角，呈逆时针方向压住第一个小三角。同样的办法，铺满一圈小三角，最后一个小三角压在第一个小三角之下。

Then fix the second triangle in a counterclockwise direction and press the first triangle. In the same way, a circle of triangles is prepared, and the last triangle is pressed under the first triangle.

制作第二圈，比第一圈圆形半径少0.5cm左右，但是尖角的间距应该相当。

Make the second circle, which is about 0.5cm smaller than the circular radius of the first circle, but the spacing of the angles should be equivalent.

完成第三圈。注意每个三角的尖部反向延长线都要对准圆心，间距一致，呈现正圆形放射状的效果。

Complete the third circle. Note that the inverted extension lines of the tips of each triangle are aligned with the center of the circle, and the spacing is equal, showing a radial shape.

13

完成第四圈。开始过渡到花蕊。

Complete the fourth circle. Start the bud part.

14

完成第五圈。越往后,缝合的面料层次会越多,难度也会增加,要注意控制好力度。

Complete the fifth circle. The more layers of fabric, the more difficult it will be, and the more attention should be paid to control the strength.

15

第六圈是花蕊部分,这一层花瓣一定是8个,所以要严格调整三角位置的对称。这里是一个八角花的造型。

The sixth circle is the part of the flower bud, which has fewer petals and must be eight. So the symmetry of the triangle should be strictly adjusted. Here is the shape of an octagonal flower.

16

最中心用一个菠萝结装饰,这样所有的缝线都隐藏在花瓣之下。尖角花部分完成。

The center is decorated with a pineapple knot so that all stitches are hidden under the petals. The sharp flower is partially completed.

三、针线盒的故事

朝鲜族有这样的民俗故事：女儿出嫁时娘家母亲要准备针线盒。针线盒里有针、线、剪刀、尺子，但出嫁当天姑娘只带装有针、线、尺子的盒，剪刀不能带走。因为剪刀代表剪掉和娘家的情，所以要等到第三天回门时再带走。针线盒里的物件也都有代表意义。尺子象征公公，要公正，用来画线，有决策权；剪刀象征婆婆，是顺着画线下剪刀，要执行；针是儿子，要带头走；线是儿媳妇，要随着走。线始终跟着针走，这个意思是姑娘嫁到婆家后，一定跟随婆家走，永远听丈夫的话。这个针线盒来历的故事，是我的姨妈告诉我的（她今年99岁），她让我永远记住做女人的道理和做母亲的责任。

三、Story of A Sewing Box

There is such a folklore in Korean: when the daughter is married, her mother prepares a sewing box. Needle, thread, scissors and ruler are included in the sewing box. But on the day of the marriage, the girl only takes away the box with needles, thread and ruler, and the scissors could not be taken away. Because the scissors represent the cut off feelings with mother home, so the bride usually takes it away when return to her mother home at the third day after marriage. The objects in the sewing box are also representative. The ruler symbolizes the father-in-law, fairness, decision-making power, used to draw lines; the scissors symbolizes the mother-in-law, implementation, to follow the drawing line to use the scissors; the needle is the son, who take the leader; the thread is the daughter-in-law, who follow the leader. The thread always follows the needle. This means that after the girl is married to her husband, she must follow her husband's family and always listen to her husband's words. The story of this sewing box comes from my aunt (she is 99 years old), she let me always remember the truth of being a woman and the responsibility of being a mother.

作品索引

- 002 百衲衣
- 004 彝族贴补绣女衣裙
- 004 土族七彩绣女长衫（复制）
- 005 苗族堆绣鸟纹女上衣（局部）
- 006 白族钱形纹拼布围兜
- 006 花苗上衣披肩（局部）
- 014 被面《蛇紫嫣红》
- 019 棉被《虹》
- 023 棉被《琴瑟和鸣》
- 026 棉被《和谐Ⅰ》
- 030 窗帘《吉祥如意Ⅰ》
- 034 窗帘《窗棂之美》
- 036 窗帘《天外天》
- 046 隔断帘
- 050 桌旗
- 054 桌旗与餐垫
- 060 桌旗《流光溢彩》
- 066 桌旗《自始至终》
- 068 茶盘垫
- 072 朝鲜族传统大衣
- 074 朝鲜族改良女衣裙
- 076 女式大衣
- 078 女式背心
- 080 朝鲜族改良女衣裙
- 082 背心裙
- 084 背心
- 086 大衣
- 088 连衣裙
- 092 背心与开衫
- 094 长裙
- 096 僧袍
- 098 孙子的"百天衣"
- 100 孙子、孙女的"出生服"
- 102 朝鲜族女童套装
- 104 孙女的衣裙
- 108 孙女的百天衣披风
- 110 披肩
- 111 围巾
- 134 挎包
- 136 手包
- 137 扇子
- 138 首饰
- 144 壁挂《百花争艳》
- 150 壁挂《百花齐放》
- 155 壁挂《念想》
- 158 壁挂《光芒四射》
- 161 壁挂《节俭之美》
- 167 壁挂《路Ⅰ》
- 172 壁挂《自然之美》
- 174 壁挂
- 180 壁挂《玫瑰花园》
- 184 包袱皮《惜福百世》
- 188 遮尘布
- 222 包袱皮《温暖的味道》

Index

- 002 Bainayi
- 004 Female Dress with Patch Embroidery of Yi Minority
- 004 Gown with Colorful Embroidery of Tu Minority (reproduced)
- 005 Miao Women's Embroidered Birdprint Blouse (part)
- 006 Money Pattern Patchwork Bib of Bai Minority
- 006 Flower Shawl of Miao Minority (part)
- 015 Quilt cover *Colorful World*
- 021 Quilt *Rainbow*
- 025 Quilt *Conjugal Bliss*
- 027 Quilt *Harmony* Ⅰ
- 030 Curtain *Good Luck* Ⅰ
- 034 Curtain *Beauty of window lattice*
- 037 Curtain *Outer heaven*
- 046 Partition curtain
- 050 Table runner
- 054 Table runner & Dish cushions
- 060 Table runner *Ambilight*
- 066 Table runner *From beginning to end*
- 068 Tea tray pads
- 072 Traditional female robe of Korean
- 074 Reformative female jacket and skirt of Korean
- 076 Female coat
- 078 Female vest
- 080 Reformative female jacket and skirt of Korean
- 082 Vest dress
- 084 Vest
- 086 Coat
- 088 Dress
- 092 Vest & Shirt
- 094 Dress
- 096 Ragged robe
- 098 Grandson's suit for one-hundred days old ("Baitianyi")
- 100 New born baby's clothing
- 102 Girl's jacket and skirt
- 104 Girl's jacket and skirt
- 108 Granddaughter's cloak for one-hundred days old
- 110 Tippet
- 111 Scarf
- 134 Bag
- 136 Handbag
- 137 Fans
- 138 Accessories
- 144 Wall-hanging *Hundreds Flowers Blossoming*
- 151 Wall-hanging *Hundreds Flowers Flourishing*
- 157 Wall-hanging *Missing*
- 159 Wall-hanging *Radiant*
- 163 Wall-hanging *Beauty of Frugality*
- 169 Wall-hanging *Road* Ⅰ
- 173 Wall-hanging *Beauty of Nature*
- 174 Wall-hanging
- 181 Wall-hanging *Rose Garden*
- 185 Bojagi *Hundreds Years of Cherish*
- 188 Cover
- 223 Bojagi *The Taste of Wormth*

后记

决定出这本作品集之前，我是一心沉浸在自己的"拼布王国"中、忘却外部世界的。这些年婉拒了很多媒体约访、出版社约稿和院校约课。这样做的原因有二：其一，我真心觉得自己还需要更加努力，尤其被大家夸赞的时候；其二，更重要的是，我也想趁着现在双眼未花，尽量抓紧一切时间创作。遨游在自己的拼布世界中是最令我快乐的事，就像我自己常说的：

"我的一生都在一针一线中。针与线的语言让我学会放慢脚步，在倾听一针一线的诉说中感受平心和静气。一根针、一根线凝聚了我一生的耐心，一针一线做出来的拼布作品背后，有我的努力和坚持、有我的创新和创意、有我的喜怒哀乐，也有我的责任心和真心的爱……"

促使我出书的原因，是一次考察——2018年3月，中国流行色协会的朱莎会长带我去杭州一带考察丝绸印染厂和丝绸服装加工工厂。每到一个企业，我都会看到很多布头们被堆集在落灰的角落，甚至被丢弃在垃圾桶中，那时的我就像是被夺走了快乐和对美的热爱一样，感到失落，难过不已。因为在我的拼布世界里，布头是犹如生命一样宝贵的存在，每一块小小的布头都是上帝赋予我的最好的礼物。于是，我下定决心要出这本作品集，我要让大家看到我用布头所做的拼布作品，从中感受到布头的魅力——它们应该被珍视，而不该被浪费，它们是对社会和他人有益的！

一块块布头经过我们的双手，会变成一幅幅美好的艺术品，从而给他人带来由衷的喜悦和幸福感，这其中有引导向善的作用和力量、也有环保节俭美德的启示，更是一种态度——对社会和生活负责的态度。

如果大家喜欢我的作品和故事，请一定要把这样的理念融入自己的生活和创作中。认真对待每一块布头，认真对待身边的美，就像认真对待每一段人生、每一日的分分秒秒一样。

此外，出版此书还有一个很私人的原因。"此生我唯一的遗憾是，没有为你写一本书"——这是21年前，丈夫因病去世临终前曾对我说过的话。今天，他的愿望已然实现，只是他可能猜不到，这是一本用拼布写成的书。

感恩拼布！是拼布，伴我度过与丈夫并肩面对疾病的苦难时光，填满了亲人永别后漫长的月下孤独。丈夫说："有它陪你，我安心。"当然，要让他安心。自他走后，我选择嫁给拼布……

感谢王琪先生！感谢为这本书出版付出心血的王云、刘琦、沈飞、刘婷、唐广丰，感谢北京服装学院金媛善拼布研习班的学员们！感谢我的家人，感谢精神和经济上始终支持我的儿子……

金媛善
2019年8月于哈尔滨

Epilogue

Before making up my mind to publish this collection, I was immersed in my own "patchwork kingdom" and forgot the outside world. Over the years, I have politely rejected many media interviews, press releases and college appointments. I have got two reasons. Firstly, I really feel that I am not good enough, especially when I am praised by everyone, I feel that I need to work harder. Secondly, and more importantly, I also want to spend my time making new pieces before my eyes cannot see. I am traveling in my own patchwork world. This is the most fun thing for me, as I always say,

"My whole life is spent in stitches and threads. The language of the needle and thread has always slowed me down. I feel calm and peace in listening to a needle and a thread. They have condensed my life. Behind the patchwork, there lies my hard work and persistence, my innovation and creativity, my joys and sorrows, my responsibility and sincere love..."

The reason that prompted me to publish the book was an inspection. In March of 2018, President Zhu Sha of the China Fashion & Color Association took me to Hangzhou to inspect the silk printing and dyeing factory and the silk garment processing factory. Every time I went to a company, I saw that many cloths were piled up in the corners of the ash, or even discarded in the trash. At that time, I was like being taken away from happiness and love for beauty. I feel lost and sad. Because in the world of my patchwork, the cloth is as precious as life and every little piece of cloth is the best gift that God has given to me. So, I made up my mind to make this collection. I want everyone to see the patchwork I made with the cloths, and I feel the charm of the cloths — they should be cherished, not be wasted, and they can be good to the society and other people.

A piece of cloths passes through our hands and turns into a beautiful piece of art, which brings joy and happiness to others. This has the inspiration and strength to guide the good, and the enlightenment and virtue of environmental protection. It is an attitude — a responsible attitude towards society and life.

If you like my works and stories, please be sure to incorporate such ideas into your life and creation. Treat each piece of cloth seriously and take the beauty around you seriously, just like taking every piece of life and every day seriously.

Besides, the publication of this book has another personal reason. "My only regret in this life is that I didn't write a book for you", my husband said to me before he passed away 21 years ago. Now, his wish has been fulfilled, but he may not figure out, this is a book written in patchwork.

Thank you, patchwork. It is patchwork accompanied me by the time of suffering with my husband to face the disease, filling the long lonely month after my beloved one's decease. My husband said, "the patchwork can accompany you, I am relieved". I want to reassure him. Since his death, I choose to "marry" to patchwork...

Thanks to Mr. Wang Qi. Thanks to Wang Yun, Liu Qi, Shen Fei, Liu Ting, Tang Guangfeng for your devotion. And thanks to the members of Jin Yuanshan patchwork class of Beijing Institute of Fashion Technology. Thank you! Thanks to my family, and my son who support me spiritually and financially...

Jin Yuanshan
August 2019 Harbin